George Anderson

I0397386

Quantum Entanglement and Synchronicity.

Force Fields, Non-Locality,
Extrasensory Perception.
The Astonishing Properties of
Quantum Physics.

Summary

"It is my personal opinion that in the science of the future reality will neither be "psychic" nor "physical" but somehow both and somehow neither...
It would be most satisfactory
if physis and psyche could be conceived
as complementary aspects of the same reality."

(Wolfgang Pauli, Nobel Prize in Physics,
in a letter of 1950 to Abraham Pais)

Introduction. What This Book Is About

For many centuries extrasensory perceptions such as telepathy, premonitions and foresight have been considered fantasies, illusions or fraudulent inventions.

This was the direct consequence of materialistic domination over science and the a priori denial of any reality that could not be verified in the laboratory.

Despite this, we have all had simple experiences such as strange coincidences, presentiments or even readings others' thoughts and intentions. That they were not illusory is proven by the fact that we have often benefited from them in our everyday lives.

Now finally, in recent decades, scientific evidence is emerging of the existence of a higher level of consciousness, a collective mind where ideas and thoughts common to all of humanity dwell: a psychic cosmos from which it is possible to draw and from which we receive signals and information.

In 1980 the *quantum entanglement* was experimentally confirmed, i. e. the property of elementary particles to communicate with each other without limits of space and time, in a dimension that is not subject to the known laws of physics and can be compared to a universal mind.

The Global Consciousness Project experiments, conducted at Princeton University, have undoubtedly demonstrated the existence of a global consciousness, ready to react emotionally when major events involving humanity occur.

This Project is based on electronic equipment distributed in 41 nations on all continents, able to record the mood of human communities.

On the occasion of the terrorist attack on the Twin Towers in New York, on September 11, 2001, the University of Princeton instruments recorded a very high peak of "anguish" in the global population feelings: the amazing fact is that the emotional peak was not recorded after, but two hours before the event happened.

This book talks about all the confirmations acquired in the last five decades to the theory of the *Anima mundi* dear to the Greek philosopher Plato, and then to the intuition of the *collective unconscious* of the well-known psychotherapist Carl Gustav Jung, until the definitive confirmation, coming from eminent scientists and Nobel prizes of what was predicted by quantum physics, that is the existence of a non-local level where particles, even if separated by immense distances, know everything about each other and behave as if they were one.

This book is neither a scientific text nor a philosophical or a para-religious one. The author is a popularizer with years of experience to his credit, able to identify the salient points of even very complex topics, managing to re-elaborate them in order to make them comprehensible to the general public.

The message of this book is that the dividing barrier between matter and psyche is collapsing, indeed, it has already collapsed.

From a universe totally based on matter aggregated by chance, humanity is definitely navigating towards a new way of understanding reality, where matter and psyche coexist and integrate.

While classical physics remains dominant in the world perceived by our senses, new levels of reality are opening up.

At a quantum level, classical physics is no longer valid: matter cannot perform its function on its own, but it needs a psychic dimension, that is, a further level, the level of non-locality. Here the whole universe becomes one, made of energy and information, and is coordinated by a harmony force without which only chaos would exist.

In the most hidden levels of reality matter cannot do without the psyche but, reciprocally, the psyche cannot exist without a matter through which to express itself.

This awareness is accompanying humanity towards a new evolutionary leap, beyond which the materialistic dominance will cease.

There will be an era of collaboration between psyche and matter, in which even phenomena that are now discussed or denied, such as extrasensory perceptions, will become a patrimony of common use in everyday life.

The Editorial Board

Part I. Essential Preconditions

The person who follows the crowd
will usually go no further than the crowd.
(Albert Einstein)

1. The Massacre of Baruhill Street

On October 27, 1992, a terrible tragedy occurred in Terrigal, a residential area of Bateau Bay on the Central Coast, not far from the populous Australian city of Sydney.

In the throes of a raptus, a man broke into Thomas Gannan's home on Baruhill Street, killing Thomas and his two daughters, 23-year-old Kerry and 18-year-old Lisa. He also killed his own 27-year-old son David, and two other people who were in the house.

The name of the perpetrator of the massacre was Malcolm Baker. According to the local newspaper *The Sidney Morning Herald*, which dealt with the matter for a long time, it all happened for futile reasons.

Certainly the Baruhill Street massacre is one of those episodes that cry out for revenge in front of the sky. As incredulous spectators, we wonder why such a heinous crime could have occurred. Believers can ask themselves: Why does God allow this?

The pain of the whole community was great. A third daughter of Thomas, Julie, who at the time of the massacre was 17 years old, wrote a heartbreaking love poem to the lost loved ones, from which I report some verses:

> *... now that everything has happened*
> *I wish it were a dream,*
> *but unfortunately it is reality,*

17

even if I can't believe it.
Dear Dad, even if you are gone
I always love you..."

I remembered this episode to make sense of the book.

Of course, when faced with the question of why these things can happen, we have no answers. However, we can imagine that if "someone" had issued a warning, it would not have happened.

Many times in the presence of sudden misfortunes of various kinds, such as natural disasters, there are people who claim to have had premonitions thanks to which they have been able to avoid the worst. Why doesn't it always happen? Why doesn't it happen for everyone? Why didn't it happen with the Gannan family?

A Presentiment not Understood

Still, it is possible that that family received a warning. As we have seen from Julie's poignant poem, the Gannan girls loved to write and eighteen-year-old Lisa also enjoyed writing verses.

Obviously a girl full of life, cheerful, sunny, what else could have written if not hymns to joy? Moreover, *The Sidney Morning Herald* reports that Lisa, although still a student, was pregnant.

So, a teacher from the High School attended by the young woman was unpleasantly surprised when, shortly before the massacre, she read these verses in her diary:

Figure 1. On the right, Lisa Gannan. On the left, her sister Kerry, also a victim killed by the lunatic, with her mother absent at the time of the massacre.

Why did you come to cry before my grave?

I'm not in the grave, I'm not sleeping.

I am a wandering star in the darkness of the night sky.

Don't cry here anymore,

I'm not in the grave, I live in the dark sky. "

When the teacher asked her why her verses were so full of sadness, Lisa replied that she did not know, they had come to her in a spur of the moment.

With hindsight, we can imagine that those verses could represent the premonition of a fatal event. But, in all honesty, we must also ask ourselves whether it would have been possible for Lisa to understand them. Lisa was probably warned, but with an unknown language, hermetic, difficult to interpret. A language that made no sense to her.

All the things that happen around us, the strange coincidences, the "signs", to be understood must make sense. The sense that we can give to an insignificant episode transforms it into a significant coincidence, that is, a synchronicity, as the psychologist Gustav Jung, who dedicated a large part of his life to the study of these topics, defines it.

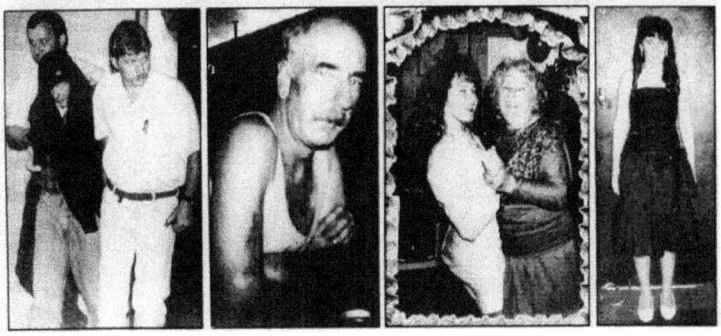

Figure 2. The Sydney Morning Herald's front page with the news of the massacre.

All of us receive messages in the form of elusive signs, strange coincidences, combinations and circumstances that are often repeated, parallelisms, but which slide on us both because we consider them the work of chance and because we cannot understand their meaning.

It is hard to believe that any entity cares so much about us and tries to send us warnings. Why should he do it? And even if he wanted to, why not express himself more clearly?

An Opposite Case

Contrary to the difficulties in interpreting the previous episode, Mr. Woods, a senior partner of a New York law firm, was much more attentive to premonitions and managed to avoid several times the imminent dangers. In fact, according to a widespread reconstruction on the net, he escaped three attacks.

The first time, on December 21, 1988, when he was supposed to be travelling on Pan Am flight 103. The aircraft, a Boeing 747-121 called Clipper Maid of the Seas, which was on its way from London to New York, exploded in flight for the detonation of a plastic explosive charge. The tragedy occurred over the town of Lockerbie, Scotland. 270 people died in the plane crash, 259 on board the plane and 11 on the ground struck by debris. Mr. Woods saved himself, because he canceled the departure.

The second time, on February 26, 1993, he escaped a terrorist attack, the one against the World Trade Center in Manhattan, consisting in the detonation of a van loaded with explosives placed in the underground parking lot. There were 6 dead and over 1,000 injured.

Finally, on 11 September 2001, he moved away from the Twin Towers shortly before the impact of the planes that caused the collapse.

Even if this story is not documented in a flawless way, there are dozens more similar, relating to people born really lucky, or . . . very foresighted and attentive to premonitions.

2. How Much Does the Soul Weigh?

There are many things around us that we do not understand. Is this an eternal condemnation, or can we hope that sooner or later we will find a way to shed light?

Everything suggests that many human capacities are not known, indeed, there are those that we do not even imagine. Others are only intuited, yet others, such as premonitions and telepathy, we would swear that they exist but we do not know how to interpret them correctly.

In the last centuries, scientific prejudices about the reality of the universe have denied any possibility of involving psychic realities in the functioning of nature.

The world around us is considered as a mass of matter subjected to mechanistic rules. According to this conception, only what can be weighed and measured is normal, everything else is pure illusion, as if it does not exist.

More Precise Scales Are Needed

With the advent of the Enlightenment, a cultural and philosophical movement that developed in Europe around the eighteenth century, it was stated that if the soul existed it had to have a weight, because a soul only spiritual was completely excluded.

So, it was thought to weigh the dying just before and shortly after death, to see if their weight varied. Having established that it remained the same, it was concluded that the soul was completely non-existent.

I thought that this experiment was condemned ridiculous by itself, until I learned that someone wanted to repeat it in our time.

A film entitled "21 grams" was made in 2003 and is based on the experiments conducted in 1901 by Dr Duncan MacDougall, in Dorchester. There are many cities with this name, but here we refer to the one located in Massachusetts (USA).

The scholar intended to demonstrate that the human soul has a mass and is therefore measurable, using scales more precise than those in use in 1700.

So, he weighed six dying patients before, after, and at the very moment of death. His conclusion was that the soul weighs 21 grams, and here comes the title of the film.

For added safety, the doctor also weighed a dog with the same method, verifying that in that case the weight remained the same. He concluded that only human beings have a soul, not animals.

Everything To Be Redone

As long as science claims to have a materialistic approach of this kind, it will never go anywhere in the dialogue between science and psyche.

What we want to affirm not only in this book, is that reality is not composed of matter alone.

The questions we ask ourselves have been the same for centuries:

- Is it possible that the psyche can also find its place in a universe made of matter?

- Is there a relationship between psyche and matter, and what are its bases?

- Does our personal consciousness live closed in itself or does it have the possibility of communicating with all the other consciousnesses in the universe?

- And what are the other consciences?

- By what means can this communication between consciences take place?

- And finally, is there a global consciousness, a "super-consciousness" that connects and unifies all personal consciousnesses?

As complex as these issues may seem, it is not impossible to approach their understanding. Many are working to facilitate the task for us and for themselves. They do it seriously and scientifically.

In this book I present the discoveries, the experiments and thre experiences conducted by the most famous scholars of the subjects involved in this research, from psychology to quantum physics.

Discussing further these topics, rather than dismissing them, means contributing to an objective understanding of our minds potential and how they manage to dialogue with the Mind of the universe.

In my role as a popularizer, I will try to expose all the hypotheses, without supporting any and without excluding any (I hope) of those based on researches carried out seriously.

I will avoid, however, those based on cheap speculations of an oriental or para-religious philosophical type, or new-age, which in my opinion are numerous and aim only to collect consensus and money (the ingredients of power) at the expense of people deluded by their persuasive arguments.

I will definitely dispute only one theory, that of absolute materialism or a priori denialism.

3 - Normal and Paranormal

The claim that psychic phenomena are "paranormal" is highly questionable. It can be defined as something out of the ordinary that does not happen often or does not happen to most people, but we all know, even from personal experience, that extrasensory perceptions occur normally and frequently.

However, for this category of events, it was decided to coin the suffix "para" attributing it a substantially negative meaning. "Para" defines all things that, according to official science, cannot be weighed and measured in the laboratory and therefore do not exist.

Following this criterion, many topics mentioned in this book would be paranormal, practically hoax or fruit of suggestion.

The opinion of many other scientists more open minded is that, instead, these are normal events, even very normal, as they affect and involve very large numbers of people, often repeatedly.

Therefore, I will also call them phenomena, but not in the sense of freak show. With the term phenomena we will refer to unusual facts that need further study in order to become comprehensible.

It is not unusual to have a presentiment that then becomes reality, or to think of a person we had long since lost sight of and receive a phone call shortly afterwards.

It is also common to share with loved ones some awareness that we had left to grow in our hearts and that we had not previously expressed.

Practically all of us have sometimes had the sensation of being watched, and turning around we have verified that someone was really looking at us.

For the proponents of materialism these are simple cases or, at most, illusions matured in the brain involutions.

According to them, the capacity to think is based on the exchange of electrochemical signals within the brain, and nothing can come out of the cranial box.

Therefore, the possibility that thoughts come out of the head or come from other heads is completely excluded. Everything we imagine is born and dies in our brain, it has no chance of going out, traveling in space and time and get to interest other brains.

Perhaps, some say, it would be possible if we connected with a cable two brains close to each other. Or a wi-fi system capable of exchanging signals between brains should be designed. In these cases everything would be based on invisible but still physical activities.

According to them, a wi-fi system based on telepathy, i. e. the exchange of pure thought, without physical connections, is completely inconceivable. Even if telepathic phenomena seem to occur, in fact (according to materialistic science) they do not occur and should be catalogued as cases, coincidences, illusions.

At this point, it is worth wondering what science is.

Is it a system in which everything proceeds according to the prejudices of the majority, and according to the unshakable "ipse dixit" of some sacred monsters of atheism, or is it a system in which research is carried out without excluding any possibility, considering that too many infallible masters have had to rethink their convictions, after having imposed them on anyone without the possibility of contradictory?

During Enlightenment and until recent years, when finally the quantum physics revelations have brought down many beliefs, materialism was the only possible reality.

From the quantum physics advent onwards, a small minority of enlightened minds, in ever increasing numbers as scientific evidence made materialistic theories obsolete, have opened new horizons by re-evaluating the concept of the psyche as essential component, in the same way as matter, in the description of reality.

The *British Society for Psychical Research* was the first organization ever to be established for the scientific investigation of claims of psychic phenomena.

British Society for Psychical Research

This Society, (abbreviation: SPR, *Society for Psychical Research*) was founded in 1882 by three members of Trinity College of Cambridge: Edmund Gurney, Frederic William Henry Myers and Henry Sidgwick, and Edmund Rogers.

It is a non-profit-making enterprise founded in order to investigate "*that large group of debateable phenomena designated by such terms as mesmeric, psychical and Spiritualistic*", and "*to approach these various problems [in] the same spirit of exact and unimpassioned inquiry as has enabled science to solve so many problems*".

Therefore, Society was born precisely to do good research on parapsychological or paranormal phenomena, and to evaluate what was true.

The research was mainly aimed at six areas: telepathy, mesmerism or magnetism, mediumship, apparitions, physical phenomena that occurred during seances and, finally, the history of all these phenomena.

Currently the society is located in London and in Cambridge. A French branch of the SPR was founded in 1885 as *Société Francaise pour les Recherches Psychiques* (SFRP). Subsequently, the *American Society for Psychical Research* (ASPR) was founded.

Among the supporters of the society there were Gustav Jung, Mark Twain, Lewis Carroll, Carl Alfred, Lord Tennyson, J. B. Rhine and Arthur Conan Doyle.

Among its presidents we can mention, from 1996 to 1999, the Italian David Fontana, professor of Educational Psychology at the University of Wales in Cardiff.

The association's activities continue with the publication of the quarterly journal *Journal of the Society for Psychical Research* (JSPR).

Still not Enough

The reception of official science was cold. Hermann Ludwig Ferdinand von Helmholtz, a German physician, physiologist and physicist, one of the most versatile scientists of his time dubbed "Chancellor of Physics", declared that never, nor if he had listened to the testimony of all the members of the SPR, nor if he had seen it with his own eyes, could have believed the transmission of thought among different people.

We could remember that similar things were said for the transmission of voice and images until the radio waves were discovered.

Perhaps the discovery of "psychic waves", currently unknown in ordinary physics, will make many people think again. As we will see, at the quantum level, similar "waves" are already being investigated.

Unfortunately, science is very conformist and everyone adapts to the existing situation, especially if it has a lot to lose. The situation in the research world is such that serious studies on extrasensory perceptions are still very few. In fact, if someone wanted to turn his interest to the paranormal fiel, he would certainly not find funding and would see his career seriously compromised.

Fortunately, things are changing, albeit slowly. Denialist materialism is now an obsolete doctrine. There are still, however, groups stubbornly committed to defending it, like the last Japanese barricaded in their forts who do not want to accept the reality of a war now lost.

4- . The Walls of Jericho Creak

There are some recent discoveries, as unexpected as they are documented and confirmed, that are upsetting the beliefs of the physics of the last centuries and can decisively affect the understanding of ESP phenomena.

The revelations in the last few years, especially in the quantum physics field and in the study of the universe, give us a completely different view of the reality in which we are immersed; I will summarize here some of them, useful for the understanding of the following chapters.

The "Junk DNA"

In molecular biology, non-coding DNA is defined as any DNA sequence not subject to RNA transcription, i. e., in practice, without any known function. The lack of apparent utility for this DNA has led to coin the term junk DNA.

Although several hypotheses have been formulated, the only thing that has been established is that about 98.5% of the human genome is composed of non-coding sequences, that is, junk DNA, *apparently* useless.

Even considering some improvement variants, the percentage of DNA that does not seem to be used remains high (around 72. 5%).

Significant similarities have been noted between base pairs of human DNA and those of fish, dogs or chickens. Apparently, there are elements of human DNA common to species millions of years away on the evolutionary path, but which have remained intact in their uselessness. But had not evolution taught us that what is useless in a species is eliminated or replaced with improvements?

It is very likely, therefore, that all this non-coding DNA should have some role that is still unclear, very different and probably more evolved than that of normal DNA.

Brain Cells Regeneration

Neurobiology, since its origins, has argued that the adult mammalian brain (including human) can not generate new nerve cells and it move towards its decay as age progresses. About twenty years ago, with the discovery of brain stem cells, this certainty was wiped out and it is now believed that the formation of new neurons in some areas of the adult brain is possible. Today we know, on the basis of experiments conducted on mice, that the brain could produce up to 10 thousand new neurons a day.

The plasticity of this essential organ is enormously greater than it was believed until a few decades ago. This leads to reconsidering the real function and capabilities of the human brain. In 1980 the prestigious scientific journal *Science* documented the case of a patient undergoing treatment at the University of Sheffield in Great Britain. On that occasion the patient, a young student, turned out to be devoid of the brain. Despite this, he had an intelligence quotient of 125, a value definitely above average. In fact, he managed to graduate in mathematics with excellent grades. *(Source: R. Lewin, "Is Your Brain Really Necessary?", Science 210, 1232-1234, 1980)*

We know many other cases of people who, despite severe cerebral impairments, manage to lead a fairly normal life. For example, it is well known the case of Michelle, a young woman from Virginia, devoid since the birth of the left half of the brain. This case has been studied and documented by Jordan Grafman, Director of the Bethesda's Cognitive Neuroscience Unit.

The CICAP publishes on its website an article stating that *"Michelle's story has been told by periodicals and newspapers as evidence of the existence of neuronal plasticity, or neurogenesis, which is often flaunted by smoke sellers; nothing has to do with Michelle's story, which instead demonstrates the adaptability of a young brain to the environment"*.It will be adaptability, the fact is that Michelle lives very well with only half a brain.

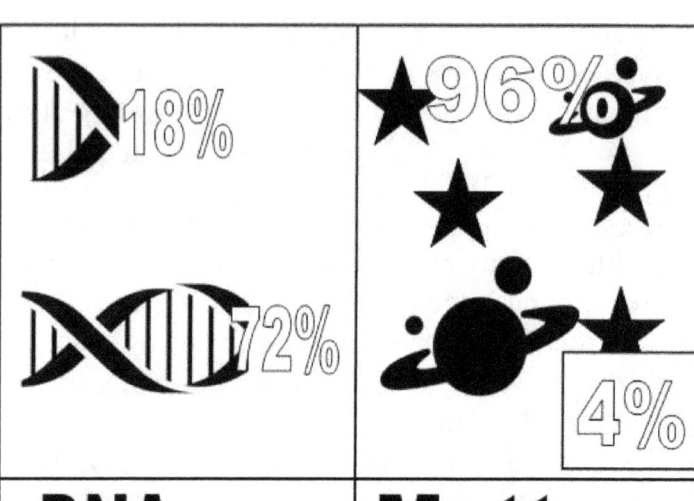

DNA	Matter
You do not know the functions of 72% of the DNA	We do not know what 98% of the matter in the universe is

What are we talking about?

Figure 3. Our knowledge of the world in which we live is still so limited that very little can be affirmed and nothing can be denied.

96% of the Universe Is Missing

Those who follow the developments of astronomical research already know that, when it was thought that everything was now known, it was discovered that 96% of the universe is made of a matter on which we know absolutely nothing, except that there is. It is not visible, it is not measurable, but we know that there is, because if it were not there the universe could not work how it works.

So far, the discoveries of the universe have been based on predictions, which have turned out to be accurate. For example, the existence of the neutrino and that of the Higgs boson were predicted. These predictions have been confirmed. Today it is predicted the existence of a part of the universe equal to 96%, and it is called "dark matter" because it is invisible and not measurable. We are waiting to know if this prediction will also be exact, or if the universe works that way for some other reason.

The statement of Lord Kelvin, one of the most eminent physicists of his time, comes back to mind. In 1900, when it seemed that Newton's mechanics and electromagnetism explained the whole of physical reality, he said: "*There is nothing new to be discovered in physics now. All that remains is more and more precise measurement*". Probably, however, in his day, not even 1% of the discoverable had been discovered.

Part II. The Soul of the World

"I simply believe that some part of the human Self or Soul is not subject to the laws of space and time."
(Carl Gustav Jung)

5. The Consciousness That Everything Is One

Since the dawn of existence, human thought has been fascinated and involved by the concept that creation is all interconnected through an overall bond, which is not material but is above matter.

Paganism and the early animist religions believed that every reality, even the simplest one, contained a spiritual presence, often connected to the soul of the whole.

Plato and the Anima Mundi

One of the first to speak of a universal soul was Plato, a Greek philosopher who lived between 428 and 340 B. C., who enunciated the concept in the Timaeus, taking it from more ancient oriental traditions. That's what he writes:

> "Wherefore, using the language of probability, we may say that the world became a living creature truly endowed with soul and intelligence by the providence of God". *(Plato, Timaeus, chapter VI, 30b-30c)*

Even Plato had his problems with materialism, represented at the time by another philosopher, Democritus, who lived from 460 to 370 B. C., who advocated the thesis of atomism. This thesis represents the first radical form of thought according to which matter constitutes the only substance and the only cause of things.

Taking up Plato's thought also the Christian philosopher Plotinus, who lived in the third century, writes in his work *The Enneads*:

> "This universe is a unique animal that contains in itself all the animals, having only one Soul in all its parts".
> *(Plotinus, The Enneads IV, 4, 32)*

Later we can distinguish between Western thought and Eastern thought.

The Anima Mundi in Western Thought

Long after Plotinus, also William of Conches, a 12th century French school philosopher, interpreting the thought of Christianity to which he was linked, spoke about the Anima mundi in these terms:

The Soul of the World is a natural energy beings for which some have only the ability to move, the other to grow, others to perceive through the senses, others to judge. The question is ... what is that energy. But, as it seems to me natural that energy is the Holy Spirit, which is a benign and divine harmony that is that from which all realities have to be, to move, to grow, the feeling, the experience, the judge."

In the 16th century the concept of the Soul of the World was supported by Giordano Bruno, who imagined God so present in nature as to affirm that nature itself was God. The philosopher Tommaso Campanella also argued that all the elements of reality have a conscience.

The Anima Mundi in Oriental Thought

In parallel to the typical forms of the West, the concept of the Soul of the World has also developed in the East, in religions such as Hinduism, Buddhism or Taoism. Even in these religions or philosophies the idea prevails that the universe is animated by a unitary and compact force. In China, this force is the Tao, the unifying activity of the cosmic dualism of Yin and Yang.

In the famous text *Tao Te Ching*, Master Lao Tse describes the Tao as follow:

> "There was something undefined and complete, coming into existence before Heaven and Earth.
> How still it was and formless,
> standing alone, and undergoing no change, reaching everywhere and in no danger. It may be regarded as the Mother of all things.
> I do not know its name, and I give it the designation of the Dao. Making an effort to give it a name I
> call it The Great."

Jung comments on this passage in his essay on synchronicity, and quotes the scholar Richard Wilhelm, who translates Tao with *"meaning"*. The Tao, therefore, represents the *meaning* of things, just like the synchronicities become such only when we attribute a sense to apparently disconnected facts. This sense is not physical, it is intuition of the psyche.

The Tao Te Ching still says:

> "The Tao (meaning) causes things
> In a misty, indistinct manner.
> So indistinct, so foggy

Images are in him,
so foggy, so indistinct
things are in him ...
....
You look for it with your eyes and you
don't see it,
this means, expressed with a name: what
is aerial.
He strains his ear and does not hear it,
this means, expressed by a name: the
subtle
He holds out his hand and does not grasp
it,
this means, with a name: the incorporeal.
...
It means form without form,
the image without a thing.
It means the nebulous-vanished,
you go towards him and you can't see his
face,
if you follow it you can't see its back."

According to Hinduism and Buddhism this force is
the Ātman, the principle of the individual and inner
Self, indissolubly united with Brahman, the principle
of the external world.

6. The Miracolates of the West Side Baptist Church

The cold evening of Wednesday, March 1, 1950, as usual, members of the choir of the West Side Baptist Church, located at 488 West Court in the town of Beatrice in Nebraska, were about to gather. At 7:30 p. m. the choir rehearsals should have started and none of the fifteen choristers used to arrive late, both out of respect for the other members of the small community and the pastor, and because the commitment was lived with the responsibility of those called to a service aimed at the glory of God.

BEATRICE DAILY SUN

"If You Didn't See It in the Sun It Didn't Happen" Member of the Associated Press

BEATRICE, NEBRASKA, SUNDAY, MARCH 5, 1950 Single Copy 5c

on Still s Gloomy r Picture

. Rail. Phone.
time Strike
uts Looming

Gas Is Blamed For Church Explosion

Find 2 Leaks In Main Not Very Distant

Gas May Have Followed Pipe Through Frozen Ground

Legal Snarls Signing Of C

Contract May Not Be Lewis Meets With

Happy Miners Toast Lewis; Rarin' To Go

Coal Diggers Elated Over Chance To Get Back To Work

Figure 4. The Beatrice Daily Sun's front page with the news of the explosion.

49

Three minutes before the start, exactly at 7:27 p.m., when usually the choir director was ready on the podium with the baton in his hand, the church exploded throwing pieces of walls and roof, furniture, furnishings, and every other part of the room, all around for tens of meters. The cause, later ascertained by the Fire Brigade, was a gas explosion.

By an incredible coincidence all fifteen choristers were saved, because just that evening no one had arrived on time.

In particular, the Reverend Pastor Walter Kiempel, his wife and daughter Marilyn Ruth had noticed, at the last moment, that the daughter's dress had a stain and they had delayed, because the other dress had to be ironed.

Herb Kipf, had to complete an urgent letter for the company where he was employed, then he went to send it. His aunt, Esther Stuermer, was late with him.

Dorothy Wood, a young student involved in assisting her father, had had problems that had been the cause of the providential delay. Since she did not live far from the church, she always walked with the neighboring friend Lucille Jones. The latter, not seeing her coming, had been waiting.

Mrs. Paul, the choir director, had to help her son with a task so that both she and her daughter Marilyn left the house late.

Harvey Ahl had also been late to take care of his children.

Figure 5. The West Side Baptist Church building after the explosion occurred on March 1, 1950.

51

Ladona Vandergrift, Royena Estes and Sadle Esh, who were travelling in the same car, had problems getting it started.

Joyce Black at first decided not to go out that night, because of the cold, but then she went anyway, but late.

Finally, Leonard Schuster helped his mother prepare a missionary meeting and did not realize he was late.

A series of circumstances had meant that none of the fifteen possible victims were affected by the explosion.

Can the case be invoked to explain this incredible chain of coincidences? In fact, they seem a bit too much to catalog them as just coincidences.

According to Carl Gustav Jung, when coincidences are so repeated they make sense and become meaningful. The study of the coincidences that are defined as meaningful led him to elaborate the theory of synchronicity.

Synchronicity indicates the link between two or more events that occur without any cause, i. e. without anyone having influenced the other, over a period of time that makes it reasonable to link them together.

In fact, between the different delays and the explosion of the church there is no causal relationship: the explosion could not have caused the delays, simply because it occurred after them.

But when all the delays become a cause of salvation for fifteen people, then we can assume that this event has special characteristics.

quale il tempo e lo spazio non influenzano gli avvenimenti né le informazioni.

If in our physical reality no one could know that the church would explode, perhaps in another level of reality it was known. We are talking about a level where time and space do not influence events or information.

In a level without any past and any future, the awareness of a danger and the occurrence of the same can be contemporary. If the consciousness were able to travel in this dimension, every danger would be known and avoidable. Unfortunately, we are not able to make this journey, but, mysteriously, sometimes the information come to look for us and we are involved in something that makes us partakers through an unknown wisdom.

Part III. Synchronicity

"Life is a game in reverse.
The truth is that all our situations or circumstances
have their beginning in our mind.
Our idea of who we are create what we become.
The great news is that you can change your ideas
and change your life. "
(Carl Gustav Jung)

7 - Carl Gustav Jung. Synchronicity and Collective Unconscious

Another way to define the Soul of the World in a more scientific way is to identify it with the *collective unconscious* theorized by Jung, a famous psychoanalyst, psychiatrist and anthropologist born in Kesswil, Switzerland, on July 26, 1875.

The collective unconscious represents a container of "ideas", the *archetypes*. It is like a higher consciousness, to which all the personal consciousnesses are connected, so that they can share its contents *(see fig. 6)*.

The archetypes contained in the collective unconscious are the forms and symbols that manifest themselves in all peoples of all cultures, and we can draw from them without having directly experienced them.

Jung builds the theory of the collective unconscious starting from the examination of the strange coincidences that affect all people. When these coincidences manifest themselves in such a significant way that they cannot be considered mere fruits of chance, they become synchronicities and need a higher state of consciousness in order to be explained.

According to the theory developed by Jung, the human psyche can be divided into three different layers: the ego, the personal unconscious and the collective unconscious. The ego contains all the notions of which one is fully aware.

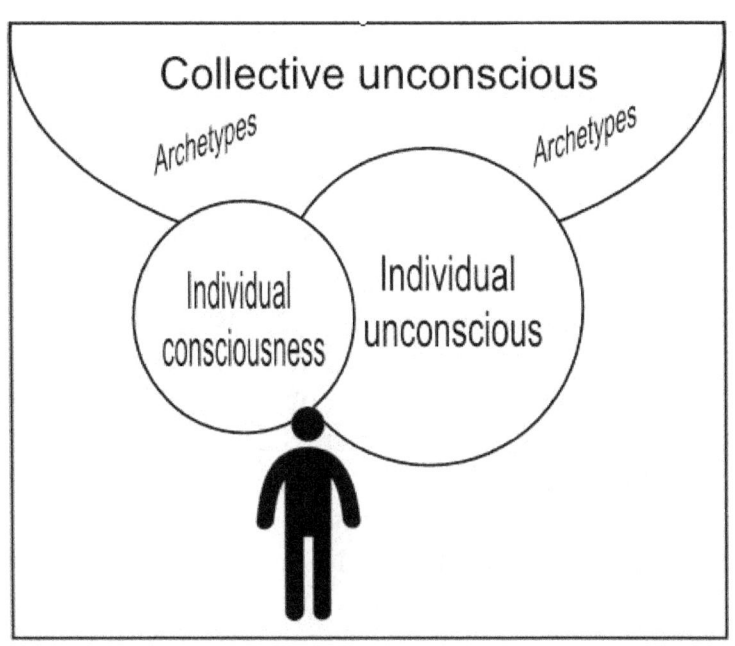

Figure 6. According to Jung, the human psyche can be divided into three layers: the ego, the personal unconscious and the collective unconscious.

The personal unconscious represents the part of the psyche where our conscious awareness resides: instincts, thoughts, emotions, behavioural patterns placed at the base of our actions. The collective unconscious contains all the knowledge of humanity, in the form of archetypes.

Synchronicity is an event through which archetypes flow from the collective unconscious to our personal consciousness, with significant effects on our reality.

In both cases reported in the previous pages, the Baruhill Street massacre involving Lisa Gannan and the explosion of the West Side Baptist Church involving the fifteen members of the choir, we can identify a synchronistic event.

A thin thread connected to their lives could have determined Lisa's salvation, and it actually determined the salvation of the choristers. We must ask ourselves: what force, what mystery, what is there at the other end of the line?

Does the content of these cases really allow them to be identified as synchronistic episodes? Let's try to examine the two events in detail, to see how they transform from simple random coincidences to synchronicity.

Synchronicity in the Baruhill Street Event

A few days before the massacre, Lisa Gannan wrote some disconcerting verses, disharmonious compared to her joyful and sunny temperament.

 **7**	Lisa Gannan writes unusually sad verses that evoke death	It is a trivial causality, due to a moment of "black humor"

 **8**	Malcom Baker enters the house of Thomas Gannan and kills six people	A case of Black chronicle how so many happen

 **9**	Lisa Gannan writes unusually sad verses that evoke death	According to what will happen shortly, we consider the fact as a premonition
	Malcom Baker enters the house of Thomas Gannan and kills six people	This event confirms Lisa Gannan's premonition

Figure 7. The fact that Lisa writes verses, as she used to do, should not seem unusual, except because these verses are unusual for her temperament.

Figure 8. The massacre caused by Malcolm Baker, during which he also kills his son, is also part of the crime news.

Figure 9. Observing the two facts together, Lisa's verses become a premonition, and the massacre is its confirmation.

A teacher from the High School she attends reads them, and asks why so much sadness. Lisa replies that she doesn't know, the verses have come out of the blue. The episode can be considered a trivial randomness *(see fig. 7)*.

Some time later, without any connection with the episode of the verses, the massacre occurs in which Lisa loses her life. Even this episode, taken in itself, is nothing more than a crime news episode as many happen every day in the world *(see fig. 8)*.

Considered each on its own behalf, the two facts have no connection: neither of them is the cause of the other. They can be considered two completely different facts.

Let us try, however, to consider the two episodes together. It is very likely that those who have lived through them will be able to connect them with a meaning, as we have done. In this case, if the two episodes are connected by a sense, they are no longer trivial, but become meaningful *(see fig. 9)*.

The facts thus linked together represent a synchronicity. In figure 10 we can see how to construct synchronicity.

Faced with the fact that Lisa was warned of the imminent danger, we can ask ourselves why she was warned, and the other victims were not.

In fact, in Lisa's case, we have discovered about the verses only thanks to the teacher, that had read them. Otherwise, we would never have known about it, because Lisa herself had not given them any importance.

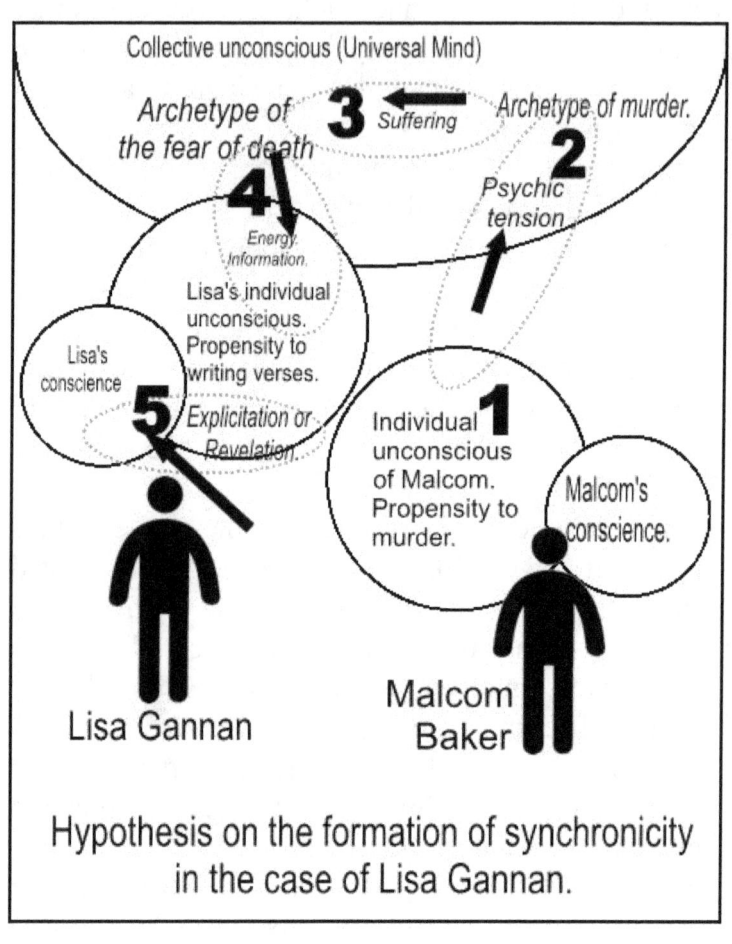

Figure 10. How the synchronicity of Lisa Gannan was formed.

Faced with such a labile clue, subject to dissolving in the memory of the receiver, we can imagine that others too may have received premonitions, but they would not have been considered worthy of attention.

Let us examine figure 10. It simply represents a hypothesis based on the theory elaborated by Jung. Obviously, this is a simplified reconstruction, aimed at understanding the involvement of the various layers of consciousness.

In point 1, Malcolm Baker's personal unconscious "swells" with murderous intent. Perhaps he, at his consciousness level, does not perceive it yet. However, in the unconscious, home to primordial instincts and emotions, the decision to commit murder is strengthening. This creates a *tension field* in the harmony of his psyche.

In point 2, the tension goes back to the collective unconscious and excites the archetype of murder.

In point 3, excitement creates a *suffering field* between the archetype of murder and the archetype of fear of death; the latter generates an *energy and information field* through which it communicates the alarm to Lisa's personal unconscious.

In point 4, Lisa's unconscious processes the information with the resources she possesses, trying to make it comprehensible to the conscience. An instrument present in her unconscious is the propensity to write verses. The information is processed using this way.

In point 5, the information is made explicit to the consciousness through the writing of verses. But Lisa does not grasp the real meaning, she just says that "they have come out of the blue".

Force Fields and Modified Reality

It is important to note that we have repeatedly used the term *"field"* to indicate a tension field, a suffering field and an information field. We will return extensively to the concept of fields in the next chapters.

Another important thing is the *transforming action* that synchronicity has exerted on Lisa's temperament, causing a sunny and joyful girl to write verses of deep sadness. The transformation is possible because the force field contains both information and energy.

Synchronicity in the Event of the Choristers Saved from the Explosion

In the case of the explosion of the West Side Baptist Church, which we will now examine, the transformation created by synchronicity will be much more evident. In fact, in this case the synchronicity will physically transform the reality causing unexpected delays in the evening of all fifteen choristers.

Here, too, if we look at the reasons for the delays one by one, we can see that these are trivial cases, as could happen to anyone at any given time. Figure 12 shows some of them.

|  | At 7.30 pm the singing tests must begin. At 7.27 the church explodes. | It is a trivial accident due to a gas leak. |

Figure 11. In itself, the explosion is nothing more than the obvious consequence of a gas leak.

These delays are by no means attributable to the explosion, that will only occur afterwards, even if only a few minutes later. The explosion is represented by itself, without any causal relationship with the delays, in the graphic above *(see fig. 11)*.

However, if we consider that *all* fifteen choristers have experienced a delay (which is absolutely unusual, since they were normally all punctual), we realize that it is no longer a matter of chance.

The coincidence of fifteen delays in the moment when the church is about to explode becomes significant, and lets us imagine, with few elements of doubt, a synchronicity has occurred *(see fig. 13)*.

In this case too we can retrace in figure 14 (by way of example) the steps that lead to the formation of the synchronistic episode.

In order to allow delays to occur, synchronicity intervened in the choristers' sensitive universe, generating the events (delays) that caused their salvation.

	Pastor Walter Kiempel, his wife and daughter are delayed to iron a dress.	It's a simple delay as it can happen at all.
	Harvey Ahl delays because he has to look after his two children.	It's a simple delay as it can happen at all.
	Ladona Vandergrift, Royena Estes and Sadle Esh are delayed due to car problems.	It's a simple delay as it can happen at all.
	Joyce Black delays because is afraid of Cold	It's a simple delay as it can happen at all.
	All the other Choristas...... delay for trivial randomness	

Figure 12. Some of the reasons for the delay. Considered individually, all the delays that saved the choristers from the explosion can be considered trivial randomness.

	Pastor Walter Kiempel, wife and daughter slow down.	Since all delay it is a SIGNIFICANT COINCIDENCE
	Harvey Ahl delays.	Since all delay it is a SIGNIFICANT COINCIDENCE
	Ladona Vandergrift, Royena Estes and Sadle Esh delay.	Since all delay it is a SIGNIFICANT COINCIDENCE
	Joyce Black delays.	Since all delay it is a SIGNIFICANT COINCIDENCE
	Herb Kipf and Aunt Ester delay.	Since all delay it is a SIGNIFICANT COINCIDENCE
	Dorothy Wood and Lucille Jones delay.	Since all delay it is a SIGNIFICANT COINCIDENCE
	Leonard Schuster delays.	Since all delay it is a SIGNIFICANT COINCIDENCE
	Mrs. Paul and her daughter Marylin are delaying.	Since all delay it is a SIGNIFICANT COINCIDENCE
	At 7.30 pm singing rehearsals must begin. At 7.27 pm a church explodes.	All these coincidences indicate that a synchronicity has occurred.

Figure 13. The fact that all the delays occurred together, in an unusual way, suggests that it is a series of significant coincidences for the connection between them and for the contemporaneity with the explosion of the church.

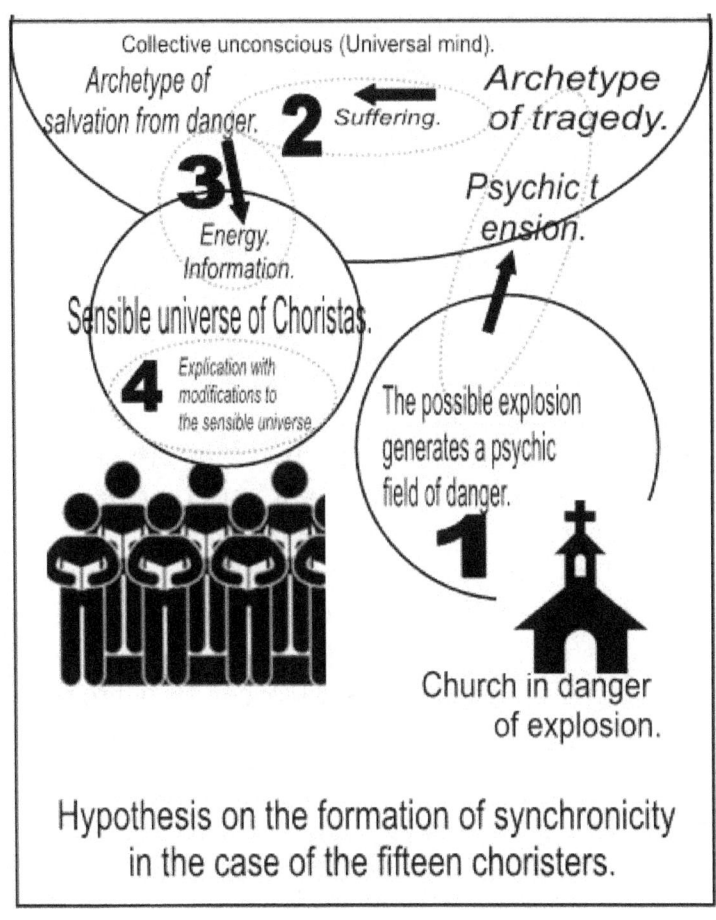

Figure 14. Assumptions about synchronicity in the case of the West Side Baptist Church explosion.

Here too we can make assumptions about synchronicity, source of salvation for the choristers.

In point 1, the possible explosion of the church generates a tension field that involves the *archetype of tragedy* in the collective unconscious of the universe. The thing to see is that the explosion has not yet occurred in our physical level of reality, but it is as if it were already present in the non-physical level of the collective unconscious, where there are no limits of time or space.

In point 2, in the collective unconscious the archetype of tragedy generates a suffering field that involves the archetype of rescue in danger.

In this case too we can retrace in figure 14 (by way of example) the steps that lead to the formation of the synchronistic episode.

In order to allow delays to occur, synchronicity intervened in the choristers' sensitive universe, generating the events (delays) that caused their salvation.

Here too we can make assumptions about synchronicity, source of salvation for the choristers.

In point 3, the archetype of rescue generates a force field directed to the choristers' sensitive universe. This force field contains not only information, but also energy. As we will see, all force fields contain energy and information.

In point 4, the force field is expressed in the choristers' sensitive universe through transformations of their real situation, by inducing in each situation a form of delay that modifies the events of the evening.

It is not unusual to observe these modifications, or transformations of the perceived sensitive universe. Jung calls them as "numinous events". The word numinous means "surrounded by an aura of sacredness, such as to cause fright and reverence together".

Among the examples cited by Jung in his works, we can take a couple of them, which include a physical synchronicity intervention in the sensitive universe of those who are living that experience.

When Reality Is Physically Influenced

A classic is definitely the episode of the beetle. In his essay *Synchronicity: An Acausal Connecting Principle*, published in 1952, Jung describes the event in this way:

> "A young woman I was treating had, at a critical moment, a dream in which she was given a golden scarab. While she was telling me this dream, I sat with my back to the closed window. Suddenly I heard a noise behind me, like a gentle tapping. I turned round and saw a flying insect knocking against the window-pane from the outside. I opened the window and

caught the creature in the air as it flew in. It was the nearest analogy to a golden scarab one finds in our latitudes, a scarabaeid beetle, the common rose-chafer (Cetonia aurata), which, contrary to its usual habits had evidently felt the urge to get into a dark room at this particular moment. I must admit that nothing like it ever happened to me before or since."

In the following, Jung explains the event in psychotherapeutic terms and describes how the episode proved to be effective for the recovery of the patient. In fact, the beetle is a classic symbol of rebirth.

On different occasion, that is in a letter written in 1945 and addressed to Prof. J. B. Rhine of Durham, USA, Jung describes a similar episode:

"I'm going to walk in the woods with a patient. She tells me the first dream of her life, a dream that left an indelible impression on her. He had a vision of a fox. He had dreamed of the animal as he walked down the stairs of her parents' house.

At that moment, not even forty meters from us, a fox comes out of the trees and runs for a couple of minutes on the path ahead of us. The animal behaves as if it shared our human condition ... "

We have to ask ourselves: was the Cetonia aurata right there, ready to hit the window glass, when the patient talked about the beetle seen in a dream?

And was the fox right there, ready to come out in front of the two walking around, while the patient was telling her dream?

The story of the two dreams generated two synchronicities that modified the physical reality of the patients and of Jung himself; they were certainly experiencing moments of pathos, deeply immersed in the narrative strengthened by the real psychic suffering of the patients. This situation, in both cases, has generated a "tension field" capable of triggering synchronicity.

As for whether the fox and the Cetonia were really there, a theory would argue that, in both cases, synchronicity had the power to move the scene to a different universe, one of the infinite possible, where the fox and the Cetonia were really present.

8. Synchronicity As Agent of Transformation of Reality

According to Jung, synchronicity works through the archetypes present in the collective unconscious. We have seen before that Jung's concept of collective unconscious is another way of defining a higher psychic reality, which since ancient times was called *Soul of the World*. In these pages we have also called it *Universal Mind*.

Now, in support of Jung's theory, we can see how all the entities referable to the Soul of the World are bringers of transformation. Every relationship between man and the Soul of the World, or whatever you want to call it, involves a change in the psychic level and often also in the physical level.

Physical Transformation. Prayer and Healing

To present an immediately understandable example of physical modification I could refer to prayers addressed to heaven for healing the sick or to obtain other graces, a practice that often gives surprising results. Every healing can be considered a synchronicity that is created between two facts disconnected in time and space, the prayer of someone located anywhere and the healing of someone else also located anywhere.

It is obvious that every healing involves an improvement, often a sensitive one, of the disease situation.

According to the theory of the collective unconscious, we can consider synchronicity as a link between the desires expressed by personal consciences and the higher sphere of the Soul of the World, from which "graces" descend.

Obviously, for those who identify the Soul of the world with *"God"*, graces come from God, but theology teaches us that in his interventions God always prefers to use means already present in creation, without resorting to supernatural ones.

In this regard, we can read a passage from a sermon by Pastor Marco deFelice (God does the impossible, Acts 12). The pastor thus comments on Peter release from the prison, where he had been imprisoned by Herod order (Acts 12):

> "Suddenly an angel of the Lord appeared and a light shone in the cell. He struck Peter on the side and woke him up. "Quick, get up!" he said, and the chains fell off Peter's wrists" (*Acts, 12:7*).
>
> One thing that strikes me about this passage is that the angel tells Peter to hurry. Although God is omnipotent, and can do anything, Peter had to hurry. Certainly, God could have made the guards fall asleep, or even die, giving Peter time to go out calmly. *But usually God does not miraculously do what can be done with the natural means that God has already provided"*.

After all, when a person turns to the faith to ask for a cure, every honest worker certainly advises him or her to consult a good doctor beforehand or in the meantime. If, on the other hand, someone turns to an exorcist (Catholic) to be freed from demonic influences, he will receive first of all the invitation to consult a good psychotherapist.

Obviously healings are not the only transformations of physical reality. We have already examined in many previous cases that there is no limit to the modifications of the sensitive universe that synchronicity makes possible. Among these we can certainly include all phenomena of telekinesis.

Desire and Intention

Speaking of synchronicities as an aid to the positive solution of human needs, we considered, in the previous passage, the desire for someone to heal and the actual healing.

A more precise term for expressing the concept of "desire" is *intention*, as it is used by Jung, but also more recently by Deepak Chopra in his books.

According to Chopra, every activity in the universe is generated by intention. In the Vedic tradition, intention is a natural power because it keeps in balance all the elements and forces that allow the universe to continue evolving.

In the fourteen Vedic *Upanisad*, Indian religious and philosophical texts composed in Sanskrit since 9th-8th centuries BC, it is said:

"You are what your deep, driving desire is.

As your desire is,
so is your will.
As your will is,
so is your deed.
As your deed is,
so is your destiny".
(Brhadaranyaka Upanisad, IV, 4,5)

9. Why It Does Not Always Happen

Although attempts were made to examine the phenomenon of synchronicity in the laboratory using scientific methods, it has always proved to be elusive. A laboratory confirmation foresees that, given certain conditions, the phenomenon usually occurs, that is, it is repeatedly reproducible, with a success rate higher than the predictable statistical average.

Example: if I flip a coin, the result can be head or tail. So, I try to pitch it and guess the result of the pitch. Of course, a single launch is not enough to compile a statistic.

If I flip the coin a sufficient number of times, say 10,000, I can guess, statistically, in 50% of cases, that is 5,000 times.

If instead I guess 6,000 times, I might have some precognitive capacity, meaning that for inexplicable reasons I know the results, before they occur, in an above-average measure. If I guess 9,000 times, I have almost certain special powers. If I guess 10,000 times out of 10,000 launches, and I repeat the test more than once with the same success, everyone will be sure that I have extraordinary precognitive abilities. Of course, orthodox scientists will say that even if they cannot find it, there is definitely a trick because *there are no precognitive powers.*

The main reason why synchronicity does not occur all the time and in all cases is the lack of the necessary precondition for its occurrence, i. e. *emotion*, or *affection.*

Figure 15. A statistical evaluation of the results can be obtained by launching a coin. With a sufficiently high number of throws, the result should be guessed in 50% of cases.

Emotional Involvement

The episodes of synchronicity are most evident in the most critical moments of our lives. Normally, our conscience tends to "keep under control" our entire psychic situation, preventing the personal unconscious from taking over the government. Let us say that, normally, "we want to see clearly", so our conscience erects a wall that curbs or filters all the contributions coming from the personal and collective unconscious, which are usually symbolic, unclear or not rationally shareable.

But sometimes in life there are difficult moments that disconcert or upset us, and thus weaken the ability to control our conscience. When events become too pressing, we tend to *surrender to fate*. Jung notes that in these cases *affection* prevails over reason:

> "Each emotional state causes a modification of the conscience, modification which Pierre Janet defined as "abaissement due nivea mentale": which means that a certain shrinking of the subconscious takes place and at the same time a strenghtening of the conscience, as many profanes in the subject can easily note, especially in the presence of intense beloved ones. The tone of the subconscious increases to a certain

measure, which easily provokes a gradient from the subconscious to the conscience. As a consequence, the conscience falls under the influence of impulses and instictive contained contents".

In confirmation of this thesis, Jung cites Albertus Magnus who, in his work *De mirabilibus mundi*, draws on *Naturalia* where Avicenna argues that the human soul has a certain property (*virtus*) to change things, and that other things are subject to it, especially when it is dragged into a great excess of love or hatred, or something similar.

"If, therefore, the soul of a man falls prey to a great excess of some passion, it can be established experimentally that it (the excess) magically constrains things and changes them in the direction towards which excess tends. [...] really the emotionality (affectio) of the human soul is the main root of all things.

This happens when, due to its emotional power, it modifies the things to which it tends. This also happens because the inferior things are subject to the soul, because of its dignity. So anyone can magically influence everything, if it falls prey to a great emotional excess. "

The fact that synchronicities occur, when there is emotional involvement is confirmed by the fact that extrasensory phenomena such as telepathy and premonition are significantly more frequent among people linked by friendship or kinship, with a maximum peak between mothers and children.

The Theory of Weak Signal

When we asked ourselves why Lisa Gannan had received a premonition and the other victims did not, we considered that probably the others had also received it, but they had not been able to recognize it. In parapsychology, reference is made to the theory of the weak signal, one of the oldest to be processed. Probably there is no harmony between the signal source and our "receiving device". As in a radio device, by adjusting the tuning you can receive stations whose existence you would not even have suspected. In our conscience many "information stations" would become more decipherable, if we knew how to tune in to their wavelength.

Part IV. Force Fields, Non-Locality and Psychic Cosmos

"I know not what I may seem to the world, but as to myself, I seem to have been only like a boy playing on the sea-shore and diverting myself in now and then finding a smoother pebble or a prettier shell than ordinary, whilst the great ocean of truth lay all undiscovered before me".
(Isaac Newton, Principia)

10. Matter, Mind and Spirit

When each of us becomes aware that we are placed in a world around us that in some way conditions our existence, we ask ourselves what is the reason for our presence. The main questions are: Why do I exist? What am I doing here? How do I participate in the great comedy / farce / tragedy of life? To start answering these questions we can configure three levels of existence: the physical level, the quantum level and the non-local level.

The Physical Level of Existence

It is the material level, that is the level of the world as we see it and as we experience it with our senses. In this level we feel safe, because all things have a weight and a dimension, a beginning and an end, a before and after, that is, a precise placement in time and space. The physical level includes all the phenomena related to our experience, such as day and night, beautiful and bad weather, the birth and death of everything, the succession of seasons, physical sensations such as hunger, thirst, pain and pleasure.

The physical world is irreparably linked to mechanism and to the laws of cause and effect.

Everything happens as a result of another occurrence. By measuring the force of the event that generated it, the new event also has predictable and measurable behaviors, and in turn produces predictable and measurable events.

In 1977, the United States launched two Voyager probes to explore the external solar system. The Voyager 2 probe was able to observe Jupiter, Saturn, Uranus and Neptune. Passing by each planet, it received the force of his gravity to continue the journey to the next planet. This was because from Earth we were able to calculate exactly the action of the gravity force of the various planets on the probe, and to establish the ways to ensure that the thrusts received were precisely those necessary to continue the race to the next destinations planned.

In 1950, Alan Turing wrote this in his book *"Calculating machines and intelligence"*:

> "The displacement of a single electron by a billionth of a centimeter at a given moment, it could mean the difference between two very different events, such as the killing of a man a year after, due to an avalanche, or its salvation".

Imagine a billiard table, with a series of balls placed on the carpet *(see fig. 16)*. In our physical level, we have the absolute certainty that no ball will move if it is not pushed by another one, or at least by a player. In the figure, the ball B will certainly remain immobile until it is hit by the ball A.

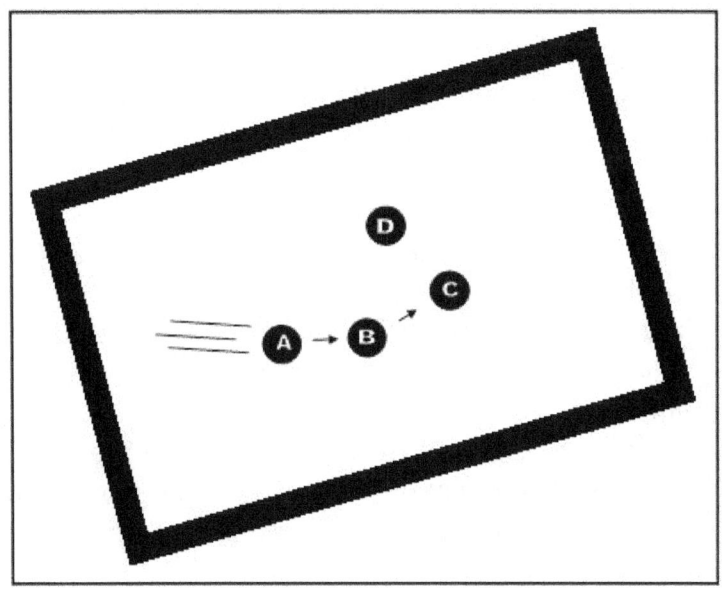

Figure 16. In a billiard table no ball moves if it is not solicited, and in this case its strength and its direction are proportional to the thrust received.

Knowing the strength and trajectory of the ball A, and the weight of the other balls, we can determine with absolute precision the force with which the ball B will move, the length and direction of its path; operations that do not require powerful computers. Every good billiard player knows how to do these calculations without any technological support. We can also predict how the other balls, eventually hit by the ball B, will move.

In 1972, Edward Lorentz titled a lecture of his: *"Does the Flap of a Butterfly's Wings in Brazil Set Off a Tornado in Texas?"*.

In today's scientific world, where everything can be weighed, measured and determined in the laboratory, a world made only of matter, placed in a dimension where time walks only forward, the answer would be yes and the determination of tornado properties would be possible, if we had sufficiently powerful computers.

The Quantum Level

At this level nothing has apparent substance, i. e. nothing can be perceived by the five senses. We cannot touch the reality of the quantum level nor can we feel its consistency with our touch, and we cannot see it, smell it or taste it.

The quantum world has a non-physical dimension that contains all the immaterial part of ourselves: our thoughts, our hopes, our desires, in practice everything that composes our ego and our self *(see fig. 17).*

Figure 17. The I or Ego represents our strictly personal sphere (facing inwards), while the Self, facing outwards, represents the entirety of ourselves in relation to the universe that surrounds us.

The *I*, better known as *Ego*, represents ourselves in relation to ourselves. The *Self* represents ourselves and our relationship with all the reality that surrounds us; the set of elements to which we refer to describe who we are.

We know that our thoughts and feelings, the spiritual relationships that bind us to our neighbor and our ambitions, are not solid, we can not hold them between our fingers. Nevertheless, we do not doubt that they exist. Yet, matter could not be if it was not animated by the energy and information contained in the quantum level.

In fact, matter itself is energy, according to Einstein's equation $E = mc2$. This equation indicates that Energy is equal to matter multiplied by the speed of light squared.

Reciprocally we can say that, as energy is mass, mass is energy. Actually, all the matter in the universe comes from the initial Big Bang, a great explosion of energy.

Since we see the mass but we do not see the energy, we are used to considering that the former is more important. Indeed, a universe entrusted to the mass alone could not function because the mass itself is inert: as we have seen in the example of billiards, mass (a ball) can move only if energy and information give it push and indicate the direction.

At the same time, we can say that energy needs matter to express itself: any energy, even if it is rich in all the information of the world, needs the collaboration of matter to manifest its potential.

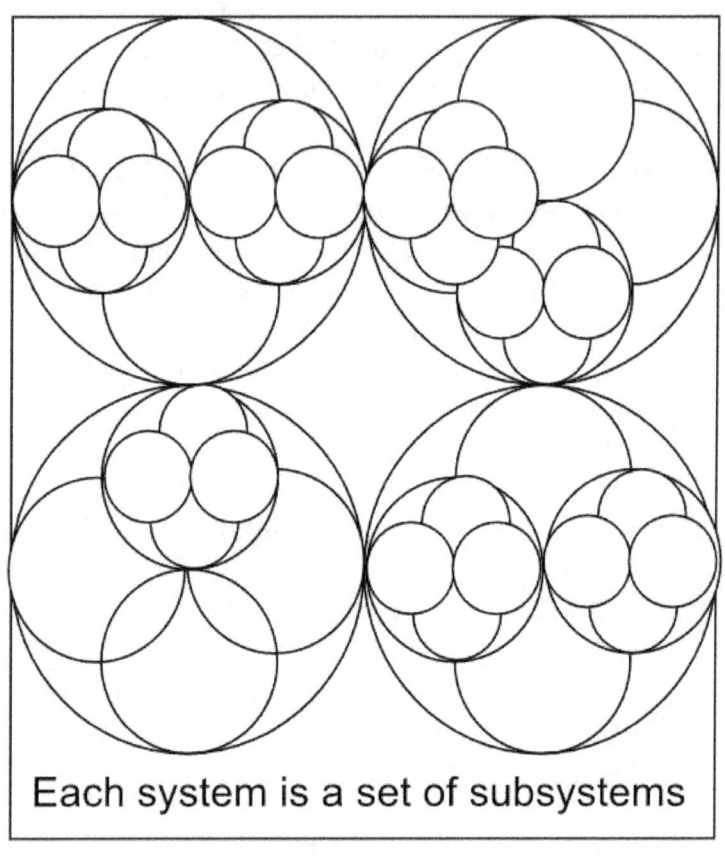

Each system is a set of subsystems

Figure 18. Each system, from infinitely large to infinitely small, is composed of subsystems. Even in particle physics, i. e.

in the infinitely small, one is not certain that one has reached the ultimate level of division.

The universe ranges from infinitely large to infinitely small. Our sensitive world is made up of matter that aggregates, from the smallest to the largest. Each body represents a complex system, composed of subsystems, i. e. smaller bodies *(see fig. 18)*.

The universe is a system made of subsystems, the galaxies.

Galaxies are made of stars and planets.

Planets are made of rocks, water, minerals and living things.

Living beings are composed of organs (skin, bones, innards, etc.).

The organs are made of cells.

Cells are made of molecules.

Molecules are made of atoms.

Atoms are made of protons, neutrons and electrons.

So far we can believe that matter, whatever its degree of aggregation, is subject to the laws of physics. The various parts of matter can communicate with each other using physical communication channels, the strength of which diminishes with increasing distance; they are also subject to the arrow of time, which moves only forward. In their movements they cannot exceed the speed of light, nor can the information exchanged by physical mediators exceed that speed. If a cell sends information to another cell, it will take some time to reach its destination and its speed cannot exceed the limit of 300,000 km per second.

Lower down, starting from the electrons ... things change.

Below are the elementary particles. For the moment, two types are known. They can be distinguished into *matter particles*, i. e. *fermions* (quarks, electrons and neutrinos, all with mass) and *force particles*, i. e. *bosons*: these particles are carriers of the fundamental forces existing in nature.

At the level of the elementary particles, all the physical laws that regulate matter are no longer valid and we are witnessing very strange behaviours, as we shall see when we talk about entanglement.

In fact, it is not even possible to determine whether the elementary particles are really particles or whether they are rather waves that vibrate.

For example, if we throw a stone into the pond, we can consider the stone as a particle, and the ripples of the water as a wave. Other examples of vibrating waves are light or music.

The conclusion is simple: the book you are reading seems to you solid, because you are looking at it with your eyes enabled to see the physical world. If, on the other hand, your eyes were enabled to see the quantum world, the book would become vibrant information, a dancing nothingness in a universe made of vibrations. You probably would not understand it is a book. As Nietsche said: *"And those who were seen dancing were thought to be insane by those who could not hear the music"*. Unfortunately, our ears are not made to perceive the music that makes the particles dance *(see fig. 19)*.

Figure 19. Elementary particles dance to the sound of instruments incomprehensible to our ears.

The quantum world is not made up of objects but of information contained in energy.

Fortunately for our survival needs, our eyes see reality at its physical level and not at the quantum level. This does not mean that quantum reality does not exist. Not only is there, but its presence has extraordinary effects on us.

We are also made up of elementary particles, that is, energy and vibrations. At a quantum level, there are no longer boundaries of space and time and our vibrations dance together with the vibrations of the whole universe: our energy shares and is shared by all the energy of the universe. In particular, our energy participates in the "intentions" of the energy that vibrates around us. How many times have we felt uncomfortable in some situation, without knowing why? Perhaps, we later learned that the other people present were divided by feelings of hatred.

If we have sometimes had feelings like "the tension could be cut with a knife", it is because the information of the negative field in which we entered was affecting our personal consciousness field.

To understand how our vibrations join, at quantum level, with those of the surrounding reality, let us try to imagine an atom, keeping in mind that we are completely made of atoms.

What do you think of the size of an atom? Of course, you knew it was very small, but did you imagine that it was also as *empty* as it appears in figure 20?

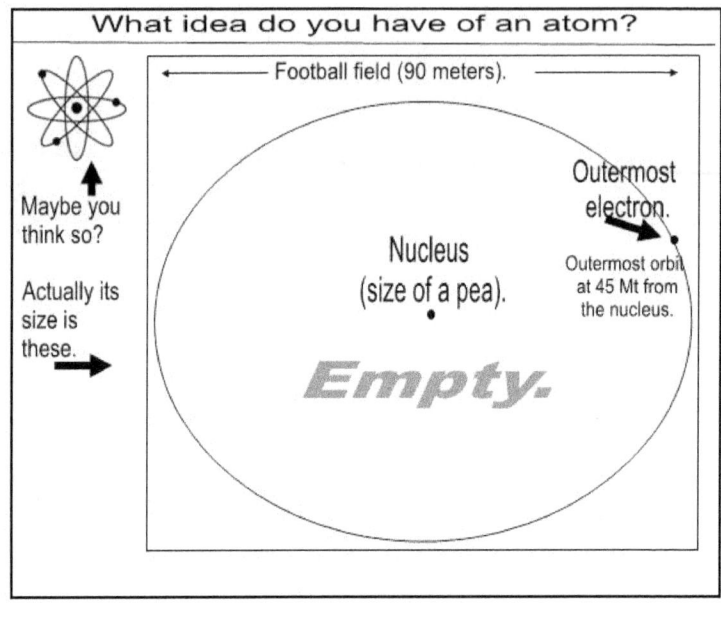

Figure 20. If the nucleus of an electron had the size of a pea placed in the center of a soccer field, its outermost electron would rotate in an orbit placed at the margins of the field. Between the electrons and the nucleus. . . huge spaces of vacuum.

Let us imagine an atom as if it had a nucleus immensely larger than reality, let us say the size of a pea, and put it in the center of a football field that, as known, measures a length of 90 meters. In this case, its outermost electron would orbit at the limits of the field. In the middle, in addition to any few other electrons, there is vacuum.

Considering this, we can understand what happens when we touch a solid object. Our impression is that the object has greater or lesser hardness, i. e. it resists the thrust of our touch. In practice, it is just a feeling of ours, because the atoms of our body and those of the object penetrate each other, just as two clouds could collide in the sky, or two puffs of smoke coming out of two nearby chimneys. This interpenetration also generates sharing: the vibrations and information of the atoms of our body spread into the body of the object touched, and vice versa. At the quantum level there is a continuous exchange of information and energy between us and the objects that surround us, and these in turn exchange energy with others.

Energy carries information with it. At a quantum level, there is a continuous exchange of energy and information. There is no longer any individuality, but a whole that shares everything.

The Non-Local Level

The third level of existence is the realm of the non-local. In this level there is neither time nor space. It is defined as non-local, because it cannot be placed in any precise place. It is neither in heaven nor on earth: it is not within us nor outside us. Simply, it is.

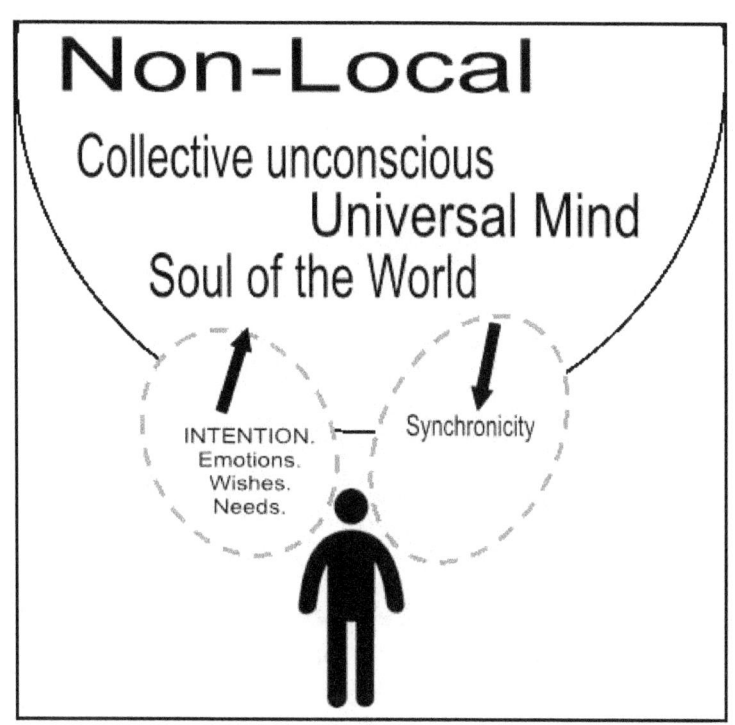

Figure 21. The non-local level, which we can call in many ways depending on studies and cultures, contains all the energy and information of the universe. We also contribute to shaping information resident in the non-local area.

It is a virtual realm that we can compare to Jung's collective unconscious. Here the information and energy of the whole universe coexist and share everything. This level is a field of pure potential, that is, the place that governs the whole universe, because it knows the whole universe *(see fig. 21)*.

There is no information, no force and no power in the universe that is not in the realm of the non-local.

The non-local level houses all the intelligence of the universe and therefore organizes and governs the whole universe. It creates and distributes the energy that shapes the physical world.

What does this have to do with us?

Simple, we are part of it. The energies, the force fields, the vibrations of which we are composed, beyond what we perceive in our physical dimension, are an integral part of the level of non-locality, they dialogue with everything it contains, they participate in the creation and distribution of the energies of the universe. We are not subjects, but participants in the non-local reality.

We can not make many observations on non-local reality. It may be useful to take up the arguments of Dr. Larry Dossey, a scholar who has published several books on the subject.

According to Dr Dossey, non-locality has three characteristics. First of all, the events that occur in the non-local *are not mediated.*

Figure 22. In the non-local level there are no limits of space or time. Reality is not bound to the laws of causality that dominate the physical level.

This means that they are not bound by causality, but each event is independent from the others, even though they can bind and coordinate with each other. Just like events that are part of a synchronicity.

Moreover, events *are not attenuated* and are *immediate*, that is, their strength remains the same (it does not attenuate) whatever the distance that separates them, both in space and in time.

We can imagine a conversation between two people, one who lived thousands of years ago in Africa and the other who lives today in the Arctic. The interview would take place as if they were both neighbours now. They would talk to each other quietly. The voice should not travel in space, everything would happen in an eternal "together" *(see fig. 22)*.

How can this be useful for us? Surely it can help us to understand that many phenomena, such as extrasensory perceptions, considered impossible by science that refers only to the physical level, become possible at the quantum level and even more at the non-local level.

11. Force Fields

Jung's synchronicity is based on events that have no causal link between them, that is, each of the events that make up a synchronicity has no relation with the others, except for the sense that is attributed to them by the observer.

This means that synchronicity does not fall within the field of official science, since it deals only with events that have a relationship between them, that is based on the principle of cause / effect. Although other scientists have suggested the possibility of the existence of other forms of connection in the universe, different from those based on causality, their ideas have been dismissed as pseudo-scientific and have not been taken into consideration.

Before the Enlightenment, great importance was given to the concepts of affinity, harmony and sympathy, useful for connecting parts of reality with non-physical links.

With the advent of the Enlightenment, these relationships were completely ignored until it became clear that invisible links and forces really existed. The most important discovery was the attraction exerted by the force of gravity. As a result of the well-known anecdote of the apple fallen on his head, Newton elaborated the three laws of dynamics.

According to the first, a body remains at rest or moves in a straight line until a force intervenes that modifies its trajectory. This force is gravity. For example, the Moon would be lost in space if it were not held in its orbit by the force of gravity exerted by the Earth. The same would happen to the Earth, if it were not subject to the gravitational influence of the Sun.

After gravity, it was the turn of magnetism, that is, the force that expresses attraction or repulsion between two bodies with a certain electrical charge.

The theory was developed by Michael Faraday, who proposed the existence of "lines of force" between a magnet and the attracted metal. It was then James Clerk Maxwell, with his equations, who demonstrated that electricity, magnetism and light are all manifestations of the same phenomenon: the electromagnetic field.

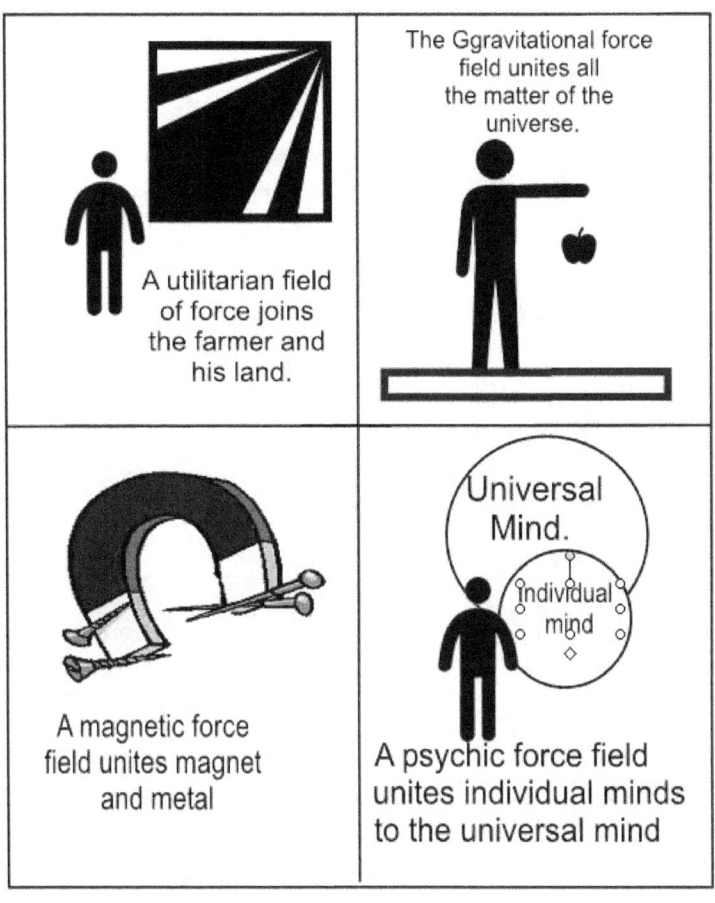

Figure 23. Representation of various forms of the force field concept.

Since then, the concept of field has become increasingly common in physics. We must make it clear, however, that this is not an exclusive concept.

The concept of field can be extended to many other relationships outside of physics, as illustrated in figure 23. To begin with, this term has always been indicative of a plot of land subject to cultivation, in the sense that the field is the place where the farmer's action is exercised.

If we accept for the term field this definition: place where the specific action of a subject is exercised, then we can conclude that the subject can be any, as much a peasant, as a physical principle such as gravity or magnetism or others, as well as a psychic principle such as synchronicity.

A utilitarian force field is created **between the farmer and his land**. We do not see it, but we do realize it exists from the fruits of the harvest.

A gravitational force field is created **between the Earth and the Moon**. We do not see it, but we realize that it exists, because the Moon does not run away into the cosmos but continues to revolve around the Earth.

A magnetic force field is created **between a magnet and a piece of metal**. We do not see it, but we realize that it exists, because the metal is attracted by the magnet.

A psychic force field is created **between the personal consciousness and the universal**

consciousness. We do not see it, but we realize that it exists, because phenomena occur that are inexplicable with the physical laws currently known.

In Jungian psychotherapy, the term *archetypal field* is often used. The difference between psychic and physical fields is that the latter are well definable and placeable in space and have their own energy. They also transmit signals that, by moving and carrying information, cannot exceed the speed of light. Conversely, psychic force fields are non-local, that is, they cannot be located precisely in time and space and they transmit information that does not have speed or time limitations. Psychic force fields are everywhere simultaneously in space and time. As we will see later, they have *quantum properties*.

The Mount of Cursing

When the Israelites, under Joshua's leadership, entered the promised land, they began their conquest by besieging and destroying all the cities of the locals. When they arrived in front of the city of Ai, Joshua thought that it was too small to engage the whole army, and he only sent a few soldiers. These, however, suffered the counterattack of the inhabitants of Ai, who pursued them in the countryside and massacred them. In despair, Joshua consulted Jaweh, who pointed out that the defeat stemmed from the defaults of the people of Israel. But then he comforted him and suggested the strategy to conquer Ai.

All Joshua's army moved close to the walls, keeping itself hidden. A squad of soldiers staged an attack and the inhabitants of Ai, thinking of repeating the previous success, went out and started chasing them in the countryside. This was enough for the other hidden troops to conquer and set fire to the defenceless city.

The victory made Joshua realize that any disobedience would cause defeats for his people, and reminded him of what Jaweh had prescribed, through Moses:

> "When the Lord your God has brought you into the land you are entering to possess, you are to proclaim on Mount Gerizim the blessings, and on Mount Ebal the curses." (*Deuteronomy 11:29*)

The two mountains are located opposite each other, on both sides of the Sichem valley, now known as the Nablus valley. The modern Arabic name of Mount Garizim, which reaches 881 meters, is Jebel at-Tur. Mount Ebal is higher, 940 meters, and corresponds to the current Gebel Eslamiyeh.

Therefore, in obedience to Jaweh, Joshua performed an epic and grandiose ceremony. All the men of the six tribes of Israel, Reuben, Gad, Asher, Zebulun, Dan, and Naphtali, went up to mount Ebal

114

for curses; the other six went up to mount Garizim for blessings.

One had to bless those who from then on would respect God's law and consequently curse those who had not done so.

The Ark of the Covenant and the Levite priests stood in the middle of the valley. When, shouting, they loudly proclaimed a blessing, the tribes placed on Mount Garizim responded with a cry that came down, echoing in the valley: *"Amen! Amen!"*. And so did the tribes on Mount Ebal, when curses were proclaimed.

This was to establish a context of blessing and curse in which all nature, from the valleys to the peaks of the mountains, was involved, and which would be perpetuated over time until the mountains themselves would remain there to bear witness to it.

We can imagine the grandeur of this event, with tens of thousands of people gathered on the two mountains. Indeed, the emotional participation of so much humanity had to create that "force field" that would be established in a dimension even greater than the written law, an eternal force, indestructible and no longer modifiable.

Bathing in the Ganges River

That was not the only occasion when force fields, generally conducive to humanity, are created by human participation.

Indian mythology has it that Ganga (hence the name of the Ganges River), daughter of the mountain king Himavan, had the power to make everything she touched pure.

That is why, according to the Hindus, the river is sacred. They believe that at least once in their lives one must bathe in the Ganges.

Many Hindu families keep a vial of Santa Ganga water in their homes. Everyone believes that this water can cleanse a person's soul of all sins, and it can also heal the sick.

Every six years believers gather at the confluence of the Ganges with two other rivers, the Yumana and the Saraswati, to celebrate the feast of the Ardh Kumbh Apple, at the end of which takes place the ceremony of the Shahi Snan, or purifying bath. An incredibly pathos-rich ceremony in which thousands and thousands of Hindus immerse themselves in the river for the soul's healing bath; flowing water is imbued with their faith and reverberates it as a healing power over the whole creation.

It is clear that this ceremony provokes emotions of immeasurable intensity even for non-believers who attend it. Surely, all participants are involved in a powerful mystical force field.

It is surprising that tens of millions of people, some suffering from infectious diseases, and even corpses of humans and animals, touch the waters of the Ganges River without anyone ever contracting diseases by bathing or drinking them.

We can find such a mystical field of healing power in many other sources of water, such as in Lourdes.

Force Fields and Archetypes

According to Jung, archetypes represent the innate and predetermined ideas of the human unconscious. Residing in the collective unconscious, archetypes are real centres of psychic energy. They manifest themselves spontaneously in the minds of individuals, especially in moments of suffering due to crises or transformations.

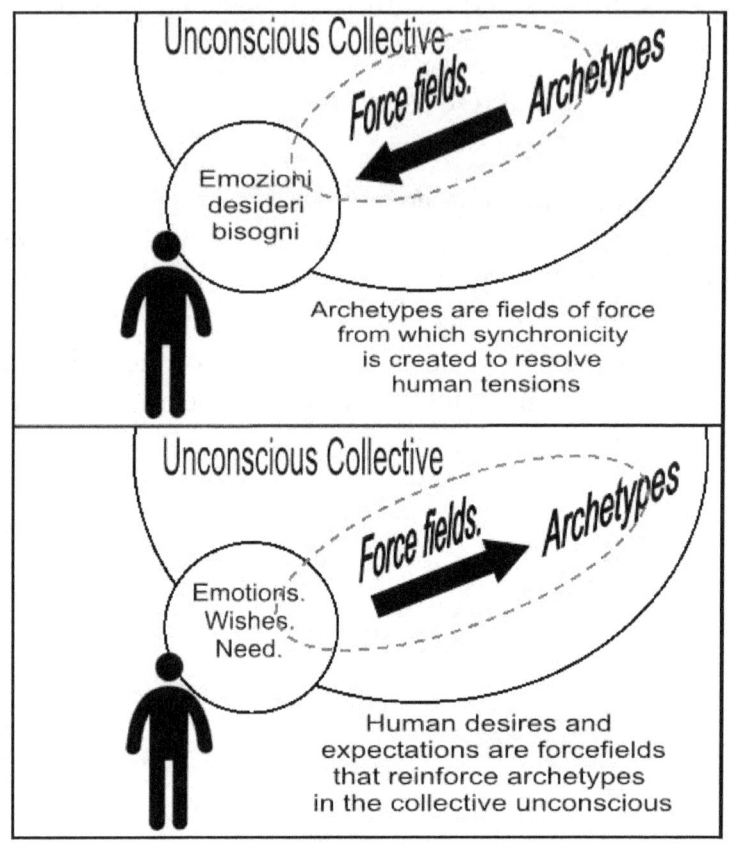

Figure 24. Reciprocity of strengthening between the intention that rises towards the non-local and the synchronicities that derive from it.

In a person the episodes of synchronicity occur thanks to the connection that is established between what his psyche gathers from the collective unconscious and what he experiences in the outside world.

Although they are innate and contain all of human history and experience, archetypes become stronger as the human species evolves. The more force fields are generated by the human psyche, the more the reference archetype is consolidated in the collective unconscious.

In this sense we could say that man's desires are shaped by that universal mind that we define as collective unconscious, but in turn the universal mind is shaped by the force fields generated by desires and expectations, that is by human intentions. Synchronicity works in both directions.

12. Mental Fields

Rupert Sheldrake is a British biologist and essayist, best known for his theory of *morphic fields* or *form fields*. The characteristic of these fields is that they have no reference to classical physics, i. e. they are not mechanistic, but are governed by unknown laws. That is what he says in his book *The Extended Mind*:

> "It was at the time when I was researching plant development at the

University of Cambridge that I came to the conclusion that living organisms are organized through fields. How do plants grow from simple embryos contained in the seeds to become gardenias, redwoods or bamboo? How do the leaves, flowers and fruits take on their characteristic shapes?

The most naive answer is to say that everything is genetically programmed. In some way each plant or animal under development follows the instructions contained in its genes. The problem with this theory is that we actually know what genes do: they encode the amino acids that make up the protein molecules.

Genes allow cells to produce the right proteins at the right times, as the body develops. But how does having the right proteins explain the shape of a flower or the structure of a mouse? No one knows. This is one of the main unsolved problems of biology.

Over the past forty years, enormous efforts have been made to study genes and control their activity. A lot of detailed information is available, but this does not mean understanding the development of a mouse or any other organism.

To say that cells, fabrics and organs simply organize themselves automatically is to say that if all the materials were delivered to a building site at the right times, the building would automatically build itself and with the right shape.

Obviously this is not the case. The buildings are not built alone, but are erected according to a project. Furthermore, the project is not contained in building materials: it is a spatial idea, a scheme of information that is lacking in materials. "

Since the last century many biologists, studying the development of living organisms, have come to the conclusion that the work of genes is not enough, and that in addition to programming the necessary materials there must exist an organizational system that oversees development. The system is believed to be based on fields, called *morphogenetics*. In the mathematical models prepared to study the development of organisms, the figure of the *attractors* has been introduced. Attractors are the objectives, that is, the purposes to which they tend and towards which they push the morphogenetic fields.

These attractors reside in unspecified spaces that mathematicians call *basins of attraction*, and have the ability to attract and direct each organism towards development appropriate to its species and genus.

Thus, the development of a mouse is shaped by the morphogenetic field of mice, and that of a daisy by the morphogenetic field of daisies.

All right, but where are these fields? How can we weigh or measure them?

This is where the absolute darkness applies. So even the mathematical models of the development of a simple daisy must resort. . . to the unknown, to explain how the daisy forms.

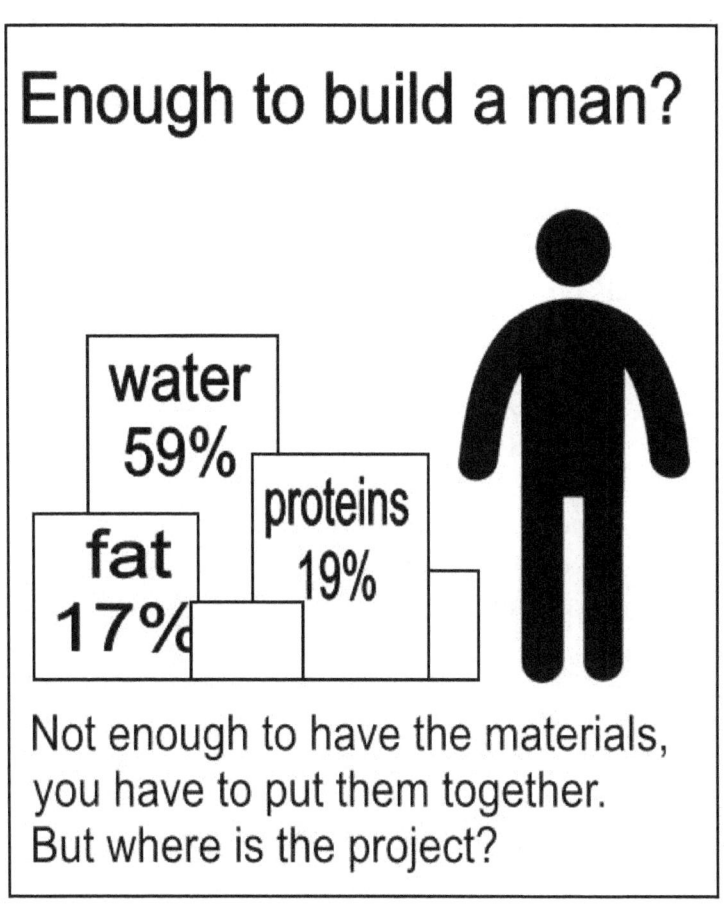

Figure 25. It is not enough to pile all the materials on site to create a building, you need a PROJECT that the materials do not contain.

It becomes mandatory to accept that in addition to physical fields such as the electric field and the magnetic field, there are non-physical and non-local fields, i. e. fields that are not measurable, and of which we do not know where they are or where they originate.

Let us consider a human being, man or woman, 30-40 years old, 1.75 meter tall and weighing 70 kg. Its chemical composition will be:

> water 59%, equal to 41.4 kg
> protein 19%, equal to 13 kg
> fat 17%, equal to 12 kg
> minerals 4%, equal to 3 kg
> 1% carbohydrates, equal to 0.6 kg
> trace vitamins 3-5 g

Do you think that by putting these elements together in a shaker and shaking at will, you would get a man? No. Even humans are not "assembled" at random, but according to the instructions of a morphogenetic field *(see fig. 25)*.

How Do Morphogenetic Fields Work?

Actually, no one knows. Rupert Sheldrake, starting from the concept that these fields are of a kind not yet recognized by physicists, suggests three properties.

The first is obvious, the morphogenetic field is a *project* that governs processes that would otherwise be random; in the absence of a morphogenetic field that guides (indeed, imposes) the development of a mouse, there would be chaos: we could have mice with eight tails and one leg, or worse mice with the body of cockroaches.

The second is that these fields, as postulated by mathematical theories, contain the attractors.

They not only know how to plan the development of the moment, but they also know how to guide what will be the future developments. Morphogenetic fields know "where they are going", they see what will happen before it happens.

They have a wisdom or, if you like, a conscience, an awareness of their task. As the project develops today, the field already sees what will happen in the future.

The third property is that the morphogenetic fields are not inscribed in the universe once and for all, but they evolve together with the same organisms they take care of. The fields of each species have a history of which they are aware *(see fig. 26)*.

Morphogenetic Fields and Morphic Resonance

A further step is to extend the operating principle of morphogenetic fields. These are limited by their own name: their function is to collaborate with the genes

for the realization of the construction projects of the
or

Morphogenetic fields	
	The vine produces grapes because its development takes place in accordance with the morphogenetic field of the vine.
	This vine growing enhances and modifies the morphogenetic field of all the vine plants.
	Pear produces pears because its development takes place according to the morphogenetic field of Pear.
	This growing pear tree enhances and modifies the morphogenetic field of all pear trees.

Figure 26. Each element of the universe, animal, vegetable or mineral, develops according to a project inscribed in its morphogenetic field.

We can imagine larger areas, in which the fields could become projects-guides to build not only the physical body, but also the character, the psychic properties.

For example, as the mouse grows, its instincts and behavior may also grow with it, according to a pre-existing project. As a young mouse its instinct will lead it to seek nourishment from its mother, when it grows up it will seek food independently, hunting or preying. In this case, we are talking about the morphic resonance between the psyche of the young mouse that is developing and that of the infinite young mice that lived before it.

Resonance is a phenomenon similar to echo, or rumble. To give a very common example, imagine turning on two microphones and approaching them; you may begin to hear whistles of increasing volume. Technically it is said that the two microphones *enter into oscillation*: each of them transmits a signal to the other and receives it of amplified return. This is what we have proposed in figure 24, arguing that archetypes and consciousness interact and strengthen.

In the same way, due to morphic resonance, the psyche of billions of mice previously lived and that of the single mouse enter into resonance, influencing each other: the individual receives all the experience of the species and the experience of the species is enriched by that of the single. We talk about the *memory of the species*.

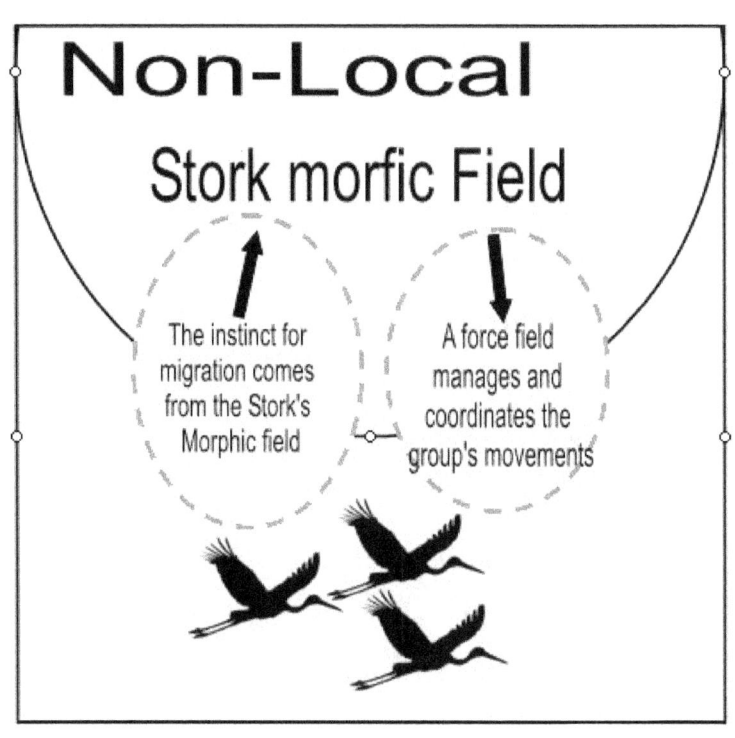

Figure 27. The movements and evolutions of schools of fish or flocks of birds are due to force fields connected with non-locality. This is why they are transmitted instantaneously to all the members of the school or the flock.

The question is, where is the memory of the species? Nowhere, we can say, because one of its properties is non-locality. Or, by analogy with what has been said so far, we could place it in the collective unconscious of mice.

I quote Sheldrake again to better clarify the concept of morphic field:

> "Other kinds of morphic fields include the behavioral fields that underlie the behavior and instincts of animals. As a kitten grows up, its instincts and behavior are shaped by morphic resonance from countless cats in the past. Its morphic fields contain a collective memory of the species. These fields interact with nervous systems and brains by imposing pattern and order on otherwise indeterminate or chaotic processes within them, as I discuss below.
>
> In addiction, the morphic fields of social groups, or social fields, coordinate the behavior of animal groups, such as termite colonies, flocks of birds, schools of fish, and packs of wolves". (*The Sense of Being Stared At: And Other Unexplained Powers of Human Minds*)

Morphic Fields

Morphic fields manifest their silent presence in our perceptions, in our thoughts and in all our mental processes. Morphic fields of mental activity are called mental fields. Through the mental fields, the mind stretches out into the environment and connects with other members of social groups.

These connections can explain telepathy, the feeling of being observed, clairvoyance and psychokinesis. They can also help to understand presentiments and premonitions, because they know how to extend into the future, since in the non-locality the future does not exist.

Force Fields in Nature. The Stones

At the beginning of this part I proposed the hypothesis of force fields that are formed due to intentions, that is of the desires and conscience of man, as in the occasion of the sacred bath in the Ganges river. However, there are natural force fields, independent of man, that man has identified since the dawn of mankind, when his ties with the forces of nature were much stronger than they are today.

Already in prehistoric times there were pilgrimage destinations, linked to the belief in the sacredness of particular places: forests, waterways, stones. Forms of worship also developed from these beliefs.

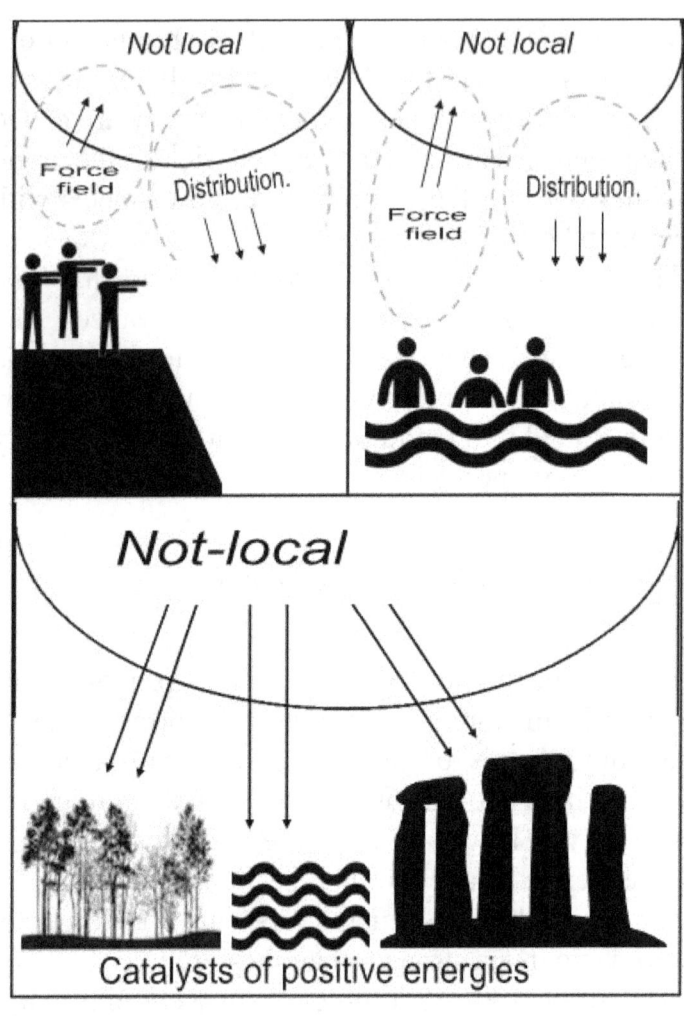

Figure 28. Men's intentions can create force fields that affect the non-local level. At the same time, many places on earth are catalysts for positive energy from non-local sources.

The earth was understood as a Mother, according to a concept present in all peoples. As a result, all things that come from Mother Earth took on a sacred meaning.

In particular, some stones were seen as containers of mystical properties. The stone, which in itself symbolizes strength and durability, became an expression of the presence of a divine, powerful and eternal power.

The stone also inspired security, since the first human dwellings were located right inside the stone, in the caves.

The fact that the caves were not only considered utilitarian shelters, but places imbued with sacredness, can be seen from the decorations in many of them, which transformed the caves more into temples than houses, such as the caves decorated with rock paintings of Lascaux or Chauvet in France, the cave of Altamira in Spain or even the Grotta del Genovese on Levanzo, Sicily.

No other material has accompanied man's history like stone. The bond begins with the construction of tools and continues today with the raising of buildings that reach the sky. We find the symbol of the stone also in the alchemical practice, aimed at the search for the Philosopher's Stone, capable of transforming lead into gold, but also of raising man from his raw nature to a higher spirituality.

In all cultures man is seen as reaching out to the sky, from which he originated. The man is only temporarily separated from the universe, given his mortal nature, but he is constantly looking for a bridge that reconnects him to the eternity from which it derives.

Often the stones become the bridge that favors this contact. In this perspective the megaliths can be seen.

Stones erected as single elements or in circles (cromlech) as in Stonehenge, England, or even in alignments as in Carnac, France, have the power to catalyze the energy of the universe *(see fig. 28)*.

This energy, coming from the whole, that is from the non-locality, generates local force fields in which man can immerse himself in order to be connected with the universal force, the divine.

The force fields generated by these places often have excellent healing properties, if for no other reason than that psychological well-being also creates physical well-being.

Pierre Teilhard de Chardin and the Noosphere

Pierre Teilhard de Chardin was a Jesuit priest, born in France, in Orcines, in 1881. Known above all as an evolutionist scientist, he developed several theories among which the one concerning the *noosphere* is of importance, a term that derives from the Greek word *nous*, meaning *mind*, and sphere, in the sense of *biosphere* or *biological sphere*.

For Teilhard de Chardin the noosphere is the collective consciousness of human beings, which arises from the interaction between all human minds.

As the organization of the human species progressed, the noosphere developed into a kind of resonance.

The concept certainly has strong analogies with Jung's collective unconscious and with the formation of Sheldrake's morphic fields.

The more humanity organizes itself by increasing its complexity, the more the noosphere takes shape. The scholar argued that the noosphere will extend, with increasing resonance and integration, until it reaches what he called the Omega Point, that is, the maximum level of complexity and consciousness towards which the universe tends in its evolution.

The publication of his writings aroused heated controversy in Catholic circles, partly because of the contrasts between conservatives and innovators in the period prior to the Second Vatican Council. In fact, Teilhard de Chardin was accused of heresy, more precisely of pantheism, the same accusation that led Giordano Bruno to the stake.

As usual, the Church has proven to be a step backwards in pursuing science, so convictions of judges as zealous as they are improvised must then be reviewed by more reflective and enlightened minds.

In fact, in recent times Teilhard de Chardin's theories have been abundantly reassessed. Pope Paul VI, in a speech on the relationship between science and faith, praised his work and pointed out him as a scientist who, in the study of matter, had managed to find the Spirit and added that his explanation of the universe manifested, did not deny "the presence of God as an Intelligent and Creator principle".

Later Cardinal Ratzinger, then Pope Benedict XVI, in his book *Principles of Catholic Theology*, admitted that one of the main documents of the Council, the encyclical *Gaudium et Spes*, was strongly permeated by the thought of the French Jesuit. He also stated that his vision was a great one, so that *in the end we will have a true cosmic liturgy*.

13. Time, Distance, Non-Locality

Before addressing the topic of entanglement, it is useful to have an idea of the dimensions in which we move. Living in the physical level, typical of the reality that surrounds us, we are used to measuring time on a scale ranging from seconds to centuries, and it seems to us that the distance between the two extremes is enormous.

Actually, a century corresponds to 876.000 hours, and therefore, considering that an hour is composed of 3.600 seconds, a century is composed of approximately 3 billion and 154.000 seconds. However, if we try to think that in astronomy time is expressed in billions of years (a billion years are ten million centuries) we can clearly see that the units of measurement commonly used are not enough.

The same can be said for distances: we ordinarily reason in terms of millimeters, meters or kilometers. However, if we want to make measurements in the galaxy, or in the nucleus of an atom, we realize that these units of measurement are not adequate.

It is enough to consult the following tables, which concern the units of measurement of time and space, to realize that our knowledge of reality is incredibly limited. We are so confident that reality consists only of what we can see and touch, that the discovery of entire immensities, that we can neither see nor measure, should lead us to an attitude of profound humility with respect to what we believe the world to be, and what it is in reality.

Classical physics and quantum physics

Sensitive reality

Atoms.

Elementary particles.

Classical physics

Matter Obeys:
-Force of gravity
-Causality
-Speed of light
-Time Arrow.

Quantum physics

Elementary particles
do not obey the rules
of classical physics .

Figure 29. Until a few years ago, we only knew about classical physics. Now it has become clear that there is another level, that one of elementary particles, where the rules of classical physics are no longer valid.

Some Distances in the Very Large and in the Very Small Expressed in Meters

Some Distances in the Very Large and in the Very Small Expressed in Meters	
Distance	Equivalent to Meters
Average distance between Earth and Andromeda (the great galaxy closest to us)	2×10^{22} 20.000.000.000.000.000.000.000 m
Diameter of the Milky Way, our galaxy	8×10^{20} 800.000.000.000.000.000.000 m
Average distance between the Earth and Proxima Centauri, the star closest to us	4×10^{16} 40.000.000.000.000.000 m
Average distance between the Earth and the	$1,5 \times 10^{11}$ 150.000.000.000 m

Sun, equal to 1 AU (astronomical unit)	
Average radius of the Earth	$6,37 \times 10^6$ 6.370.000 m
Diameter of a red blood cell	8×10^{-6} 0,000.008 m
Diameter of an atom	1×10^{-10} 0,000.000.000.1 m
Diameter of a proton, part of the nucleus of the atom	2×10^{-15} 0,000.000.000.000.002 m
Diameter of an electron	1×10^{-22} 0,000.000.000.000.000.000.000.1 m

Further deepening

The Exponential Notation

Considering the informative purposes of this book, I have deliberately refrained from inserting any element that could be difficult for those who do not have in-depth studies in the fields of mathematics or physics. I believe that many concepts can be fully understood in their general lines without the help of complex formulas. I think that an exception should be made to illustrate the concept of exponential notation, since, although it is a subject of the first school cycles, many students. . . dated like me, may have forgotten it.

The very big and the very small are likely to become incomprehensible, when they are represented by a long line of zeros, and therefore a different way of writing them is used.

For example, the number 1000 can be written as 1,000, or as 10^3. One million can be written as 1,000,000, or as 10^6. So, a billion can be written as 10^9.

The top number is called exponent. The exponent of the power of 10 is equal to the quantity of zeros that the number written in full would have.

You can also imagine the power of ten as equal to the indicator of the number of positions by which the comma, or decimal point, has been moved.

For example, 353 million, or 353,000,000 becomes 3. 53x10^8.

In the case of numbers below 1, the rule is the same. A thousandth can be written as 0.001 or 10^{-3}. Starting from 1, and moving the decimal point three positions to the left, you get 0.001.

A large molecule could be the size of 2.3 billionths of a meter. This value could be written 0.000.000.002.3 m, or 2.3x10^{-9}m.

The abbreviation method based on the power of 10 is called *exponential notation*, or *scientific notation*.

Units of Distance Measurement

Units of Distance Measurement

Name	Sm	Conversion to Meters	Type
yottameter	Ym	10^{24}m	1.000. 000. 000. 000. 000. 000. 000. 000 m	Distances between galaxies
zettameter	Hm	10^{21}m	1. 000. 000. 000. 000. 000. 000. 000 m	Diameter of a galaxy
exameter	Em	10^{18}m	1. 000. 000. 000. 000. 000. 000 m	Distances between the stars
petameter	Pm	10^{15}m	1. 000. 000. 000. 000. 000 m	Distances between the stars
terameter	Tm	10^{12}m	1. 000. 000. 000. 000. m	Distances between the stars
gigameter	Gm	10^{9} m	1.000. 000. 000 m	Distances between planets and satellites
megameter	Mm	10^{6} m	1.000. 000 m	Medium-long motorway
myriametre	*Mm*	10^{4} m	10.000 m	Diameter of a large city

kilometer	km	10^3 m	1.000 m	Size of a small airport
hectometre	hm	10^2 m	100 m	Height of a skyscraper
decameter	da m	10^1 m	10 m	Road width
meter	m	1	1m	Height of a 4-5 year old child
decimeter	dm	10^{-1}m	0,1 m	Size of the hand palm
centimeter	cm	10^{-2}m	0,01 m	Height of the back of this book
millimeter	mm	10^{-3}m	0,001 m	Thickness of a book cover
micrometer (or micron)	µm	10^{-6}m	0,000001 m	Diameter of a microbe
nanometer	nm	10^{-9}m	0,000000001 m	Size of elements in computer chips
ångström	Å	10^{-10} m	0,0000000001 m	Diameter of an atom
picometer	pm	10^{-12} m	0,000000000001 m	Diameter of the

143

				atomic nucleus
femtometre (or fermi)	fm	10^{-15} m	0,000000000000001 m	Radius of the proton
attometer	am	10^{-18} m	0,000000000000000 001 m	Size of a quark
zeptometer	zm	10^{-21} m	0,000000000000000 0 00001 m	Size of a quark
yoctometer	ym	10^{-24} m	0,000000000000000 000000001 m	Size of a neutrino

Units of Time Measurement in Seconds

Unit of Measurement	Symbol	Factor	Related Events
E−44 s	t_P	10^{-44}	Planck time, the shortest time interval that current physics can determine
yoctosecond	ys	10^{-24}	1 ys: time for a quark to emit a gluon.
zeptosecond	zs	10^{-21}	14 zs: electron life in the upper orbit in Helium-9.
attosecond	as	10^{-18}	1 as: time of the decay of an atomic nucleus
femtosecond	fs	10^{-15}	200 fs: time taken for the fastest chemical reactions
picosecond	ps	10^{-12}	4 ps: cycle time of an IBM transistor
nanosecond	ns	10^{-9}	1 ns: cycle time of a 1 GHz microprocessor.
microsecond	μs	10^{-6}	
millisecond	ms	10^{-3}	Approx. 50 to 80 ms: a blink of an eye
second	s	10^{0}	Sixtyth of a minute
Kilosecond (16,6 minutes)	ys	10^{3}	1 hour: 3600 s 1 day (24 hours): 86 400 s
megasecond (11,6 days)	Ms	10^{6}	
gigasecond (32	Gs	10^{9}	1 century: 3.16 Gs

years)			1 millennium: 31. 6 Gs
terasecond (32 000 years)	Ts	10^{12}	6 Ts: time elapsed since the appearance of Homo Sapiens
petasecond (32 million years)	Ps	10^{15}	1 Ps: Time elapsed from Oligocene to today
exasecond (32 billion years)	Es	10^{18}	
zetta second (32 billiard years)	Zs	10^{21}	
yottasecond (32 billiards years)	Ys	10^{24}	

Part V. Wonders of Quantum Physics

I have yet to see any problem,
however complicated, which,
when looked at in the right way
did not become still more complicated.
(Poul Anderson, writer)

14. Synchronicities Occur at the Right Time

In the Letter to the Hebrews, a biblical text attributed in the past to St. Paul, the author states that Christ "... has appeared once for all at the culmination of the ages to do away with sin by the sacrifice of himself" (Heb 9: 26b).

In one of his lectures, the well-known biblical exegete Card. Gianfranco Ravasi pointed out that, without prejudice to all the theological implications, the expression "at the culmination of the ages" could also be understood in its most proper meaning, that is, at the right time. In fact, he pointed out, the moment was absolutely propitious for the spread of the Gospel: the Roman Empire had studded all its territories with roads, greatly facilitating the movement of people and, with them, new ideas. The task that Jesus gave to his disciples "Go into all the world and preach the gospel to every creature" (Mk 16:15) was greatly facilitated by the world situation of the time: vast territories practically in peace and easily accessible.

If we consider that the archetypes, that is "the ideas" present in the collective unconscious of humanity, interact and are strengthened with human consciences and intentions, then it is safe to assume that the idea of a new religion is strengthened and acquired much more power, the more men address their desires and expectations to it, that is, their intentions.

Human evolution		
From four million years to 10,000 years ago		Stone Age in its various stages
From 10,000 to 200 years ago		Age of metals
From 1800		Industrial Revolution
From 1900		Computer revolution
Next evolutionary step		Science that unifies matter and psyche

Figure 30. An immense time was needed to trigger an evolutionary synchronicity that would allow the transition from the stone age to the age of metals. Subsequently, the synchronicities followed one another almost frantically.

151

Christianity spread more rapidly because of the resonance between the expectations of the new believers, constantly increasing, and the idea of Christianity residing in the non-local. The high number was a multiplier. According to the Christian faith, the "non-local" was and is the Holy Spirit, who assists and guides the disciples in their preaching.

Another example of a "right time" can be seen if we evaluate the timing of human evolution.

Human ages can be divided into Stone Age, Copper Age, Bronze Age and Iron Age. By enunciating them in this way, one does not have the perception of their duration. In fact, the Stone Age begins 3-4 million years ago, and only 10-12,000 years ago changes occur that lead to the Copper Age: 5,000 years ago to the Bronze Age, and 3,000 years ago to the Iron Age *(see fig. 30)*.

The Stone Age lasted immensely longer than the following ages. Why did man stand still so long in his early age, for more than three million years, before making the first evolutionary transition to the Copper Age? That is a question modern palaeontology cannot answer. We hazard a hypothesis: in all that long period men did not want, they did not have the intention to evolve, perhaps they did not even imagine the possibility, because they did not yet have a conscience sufficiently ready to express *intentions*.

Perhaps, they were also so few that they were unable to develop a force field that would induce synchronicities that would favour evolutionary development.

At one point, the increased number and some favorable circumstances succeeded in triggering a force field sufficient to interact with the archetype of cultural development.

From that point on, the evolution was very fast, passing quickly from the Copper Age to the Bronze Age and then to the Iron Age.

Today we live the silicon age, that is the "age of information". Today we have the communicative tools to quickly spread new ideas, able to generate force fields and expectations with respect to the frontiers of the most advanced knowledge. The fields generated today can spread and change the evolutionary path of humanity in a few decades, no longer in millions of years.

The time we are living is certainly the right time for new ideas related to the progress of quantum physics, such as entanglement, to take root and upset all beliefs about the reality of the universe. In fact, entanglement predicts a reality that is no longer made up solely of matter, but of matter and psyche. The entanglement re-evaluates and makes present all the thoughts and intuitions that humanity has felt as true for millennia, without ever being able to experience them. Entanglement begins to produce real and convincing evidence. A network of "defenders" of this new science is being created around the world, capable of generating a force field sufficient to strengthen it more and more every day. The entanglement has developed in a few years through an incredible series of synchronicities.

15. A Meeting at the Right Time

One cold afternoon in January 1932, a man with a modest appearance knocked on the door of a well-known psychotherapist's office. He decided to listen to his father's advice, concerned about his mental health, and to rely on the care of a specialist. In fact, that man was going through a very difficult time in his life. Born in Vienna at the beginning of the 1900s, he would have been in a short thirty years old, with the clear feeling of having done very little in his life. He had fallen into a serious state of mental disorder due to many tragic events involving him. In particular, four years earlier his mother had committed suicide and later he had divorced his wife, a cabaret singer, after a marriage that lasted only a few weeks.

All this led him to become alcoholic, so it often happened that late at night he was thrown out of the various places he frequented, where he was well known as a brawler. Among other things, this vice made him irascible and intractable even in the university working environment, where he had never shone for the friendliness of his relationship with others.

The psychotherapist thus describes, in his notes, his encounter with the man:

> "I had a case, a university professor, a very mono-oriented intellectual. His

unconscious had become troubled and very active; so much so that he projected himself into other men who seemed to be his enemies, and he felt terribly alone, because everyone seemed to be against him. "

The psychotherapist was Carl Gustav Jung, and the patient Wolfgang Pauli *(see fig. 31)*. From that meeting, soon strengthened by friendship due to common cultural interests, an intense collaboration developed between the two: in fact, despite his psychological problems, Pauli was a brilliant scientific mind, so much so that in 1945 he received the Nobel Prize for a discovery essential to the development of quantum physics, the well-known *Pauli Exclusion Principle.*

The meeting between the two did not happen by chance, but it was an extremely precise synchronicity for the time, and guided all humanity to the current knowledge of the universe reality.

Jung and Pauli combined the theory of archetypes and the collective unconscious of the former with the scientific rigour and severe methodology of the latter. Much of the content of this book draws on the results of their collaboration. Among other things, Pauli attributed the merit of many positive results of his studies to his dreams: he claimed to receive many useful inspirations in his dreams.

Figure 31. On the left Wolfgang Pauli, Nobel Prize for Physics in 1945. On the right Carl Gustav Jung, who formulated the theory of archetypes and the collective unconscious.

Pauli Effect

Pauli was well known among his colleagues for a very special feature.

It seems, in fact, that when he entered a research laboratory something always happened: instruments that stopped working, objects that shattered, fell to the ground or exploded.

Many physicists had come to the decision to avoid in any way to let him in, inventing the most imaginative excuses but without telling them, out of respect. One, Otto Stern, openly communicated it to him as an irrevocable decision. Certainly around Pauli a psychokinetic force field was formed, that is, capable of influencing matter.

This Pauli effect is witnessed by many of his colleagues. One of them, J. Frank, wrote to him that his equipment in Göttingen had broken down, but this time it could not be his fault because he was absent. Pauli told him that at the time of the accident he was travelling, and his train was stationary right at Göttingen station.

Despite these folkloristic considerations, Pauli was one of the main minds in quantum mechanics, along with Bohr, Heisenberg, Dirac and others, and quickly became known for his fundamental and original contributions to quantum field theory. He was the living conscience of theoretical physics.

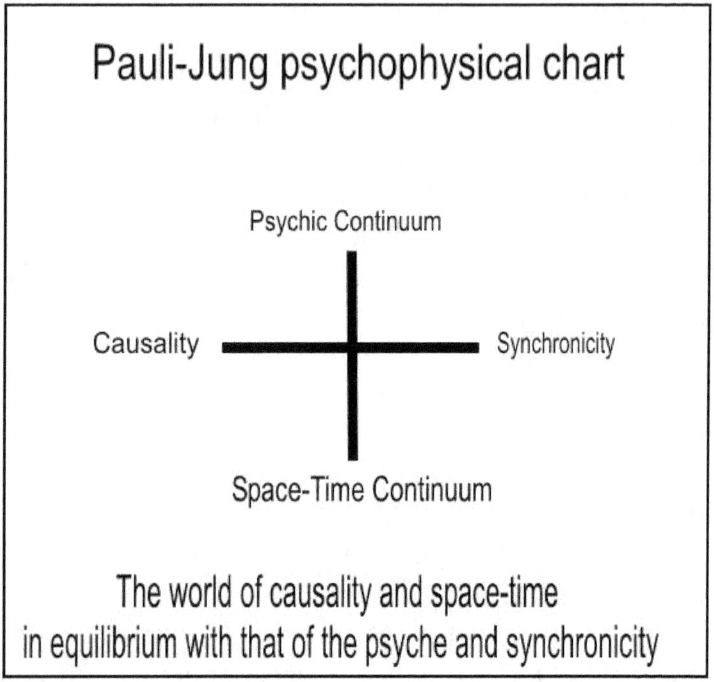

Pauli-Jung psychophysical chart

Psychic Continuum

Causality ——————— Synchronicity

Space-Time Continuum

The world of causality and space-time
in equilibrium with that of the psyche and synchronicity

*Figure 32 - Psychophysical Diagram elaborated by
Wolfgang Pauli and Carl Gustav Jung.*

His Exclusion Principle (which earned him the Nobel Prize) was the basis of the theories that allowed the experimentation of the phenomenon of quantum entanglement.

Einstein always showed him a high regard. When he was present at the Nobel Prize Award Ceremony, the creator of the theory of relativity, by then an old man, called him his spiritual son.

The Psychophysical Diagram of Pauli and Jung

The long collaboration led Jung and Pauli to elaborate the Psychophysical Diagram illustrated in figure 32.

On the left and on the right of the diagram, causality and synchronicity are balanced, assuming a collaboration between mechanistic physics (*each event is connected to a cause* that generated it) and the principle of synchronicity, which is always referred to *events absolutely not connected* by apparent causes, but only by the sense attributed to them by the observer.

The two vertical arms represent the world of the psyche, that is the world of the non-local, which balances itself with the physical world, governed by the dimensions of space and time.

This diagram confirms that reality is not made of matter alone, but of matter and psyche, and confirms that each one needs the other. A psychic reality cannot exist if it cannot be exercised on a material reality, and vice versa.

16. Entanglement

All new discoveries, especially entanglement, tend to certify the existence of a "means of mediation", that connects all the reality in which we live. The essence of entanglement is precisely this: the whole universe is interconnected through a dimension in which signals and information travel without limits of space and time. At the level of elementary particles this has been amply demonstrated. The arbitrary conclusion we can draw from it is that, since the whole universe is composed of elementary particles, the whole universe is interconnected. I say arbitrary because the entanglement, scientifically demonstrated at the level of elementary particles, has not yet been so at the macroscopic level, that is at the physical level that surrounds us, and to which we are accustomed: that of the "reality" that we can see and touch. For example, two photons were entangled, but not (yet) two tomatoes. However, experiments are being carried out, especially in the field of teleportation, which are giving flattering results.

Examining the reality from the point of view of quantum mechanics, we see that many phenomena, previously considered impossible, could actually exist. We are talking about telepathy, clairvoyance, telekinesis, and all those phenomena that pass under the title of "curious coincidences" and that the psychologist Carl Gustav Jung, together with the physicist and Nobel laureate Wolfgang Pauli, classified as synchronicity.

Someone will twist their nose, arguing that a scientifically impeccable theory such as entanglement cannot be "stained" by associating it with psychic phenomena.

Sorry, but things are just like that and the two realities are absolutely compatible. The structure of reality interpreted in the light of quantum physics and the properties of psychic phenomena may seem light years apart. Instead, they share characteristics with truly impressive similarities and compatibility.

Entanglement is a property of quantum theory certified by many experiments. Albert Einstein, who did not want to accept this novelty, because he believed that it contrasted with his theory of relativity, imagined an experiment to demonstrate its inconsistency. This well-known experiment, known as EPR from the name of the three scientists who proposed it (Einstein, Podolski, Rosen) was carried out several times in the following years, and largely confirmed the theory of entanglement, so that even Einstein's doubts turned out to be inconsistent.

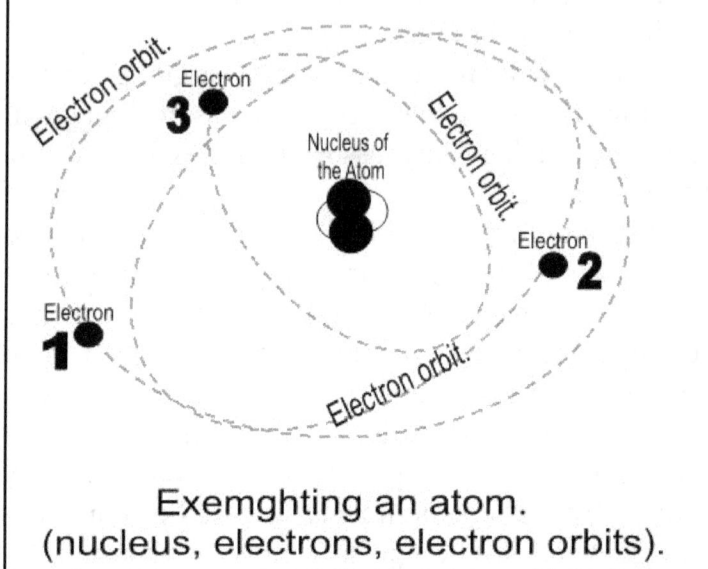

Exemghting an atom.
(nucleus, electrons, electron orbits).

The electron 1 (outermost) has an energy greater than 2 and 3, to remain in orbit around the nucleus.
Electron 2 has an energy greater than 3, but less than 1.
Electron 3 has less energy than 2 and 1

Figure 33. An electron has enough energy to keep it rotating around the nucleus. The more its orbit is external, that is, the farther it is from the nucleus, the higher the energy it possesses.

Where Entanglement Is Born. Elementary Particles.

Entanglement is a phenomenon that occurs at the level of elementary particles. A first step in understanding this is to clarify what an elementary particle is, and how we can isolate it to verify its behavior. We will refer to the category of the most known particles, the *electrons*, which belong to the category of *leptons*. As is known, electrons are very small corpuscles with a negative electric charge, which orbit around a central nucleus (composed of protons and neutrons) with a positive electric charge. Most of the electrons in the universe were created by the Big Bang.

The electron motion generates a magnetic field and is responsible for electrical conductivity and heat.

But the feature that interests us most is that the electron responds to changes in its energy and its acceleration by emitting *photons*.

The photon is a *light quantum*. To explain the concept of quantum we can make an example, and imagine going to the delicatessen to buy ham. You could ask for 100 grams, 200 or 300 grams, or 150. The grocer would cut a few slices and put them on the scale and then say: "There are 10 more grams, will I leave them?". You would pay according to weight and leave the store.

If you could go and buy photons, it would not work that way. A photon has its own weight, which is not called kilo o hectogram, but *quantum*. The photon is a *quantum*, a *unit* of light, the only possible measure, of which there are no multiples or submultiples.

You could never buy a quantum-and-a-half photon. That is, if you wanted to buy light, which is made of photons, you can not buy a quantity at will, you should measure it in whole photons.

Returning to our electron, we said that it reacts to changes in its energy by emitting photons.

Since our aim, to understand entanglement, is to make an electron emit photons, let's see how we can vary its energy.

An atom is surrounded by many electrons, depending on the element it composes. The electrons rotate around the nucleus along orbits, which also determine their energy level. The farthest ones have more energy, the ones closest to the nucleus have less *(see fig. 33)*.

Unregulated Electrons

Every now and then, for reasons that no one knows, an electron decides to break this harmony and change its orbit, that is, it decides to jump from an outer to an inner orbit.

In this initiative, however, it is strongly limited: it cannot freely make a jump, or a half jump or a long jump: it must jump exactly *one quantum*, because the orbits are arranged in a quantum order.

Not only that, assuming that our electron wants to make a jump inward to get closer to the nucleus, in the new orbit it will no longer need the energy previously possessed, so it will have to give up a part of it, exactly a quantum.

What happens to that *quantum* of energy that is gone? It becomes a particle, properly a photon, that is exactly a *light quantum*.

The fact that the electron has to jump exactly one quantum, determines a very strange phenomenon: it will never be found with the leg stretched between an orbit and the other, that is in an intermediate position. The electron will change orbit without any transition, immediately. In the electron quantum leap there is neither a space to overcome nor a time necessary to overcome it: now it is here, now it is already there and now there is a new photon *(see fig. 34)*.

We have taken a big step forward, but we need another, tiny one. To evaluate what entanglement is, we need at least two photons, because the entanglement consists precisely in the relationship between two particles, or more than two.

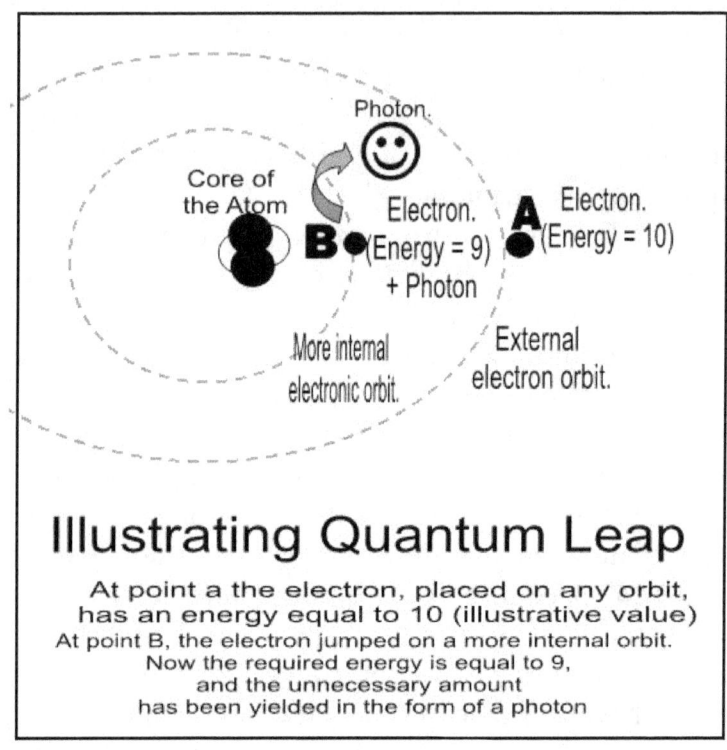

Illustrating Quantum Leap

At point a the electron, placed on any orbit,
has an energy equal to 10 (illustrative value)
At point B, the electron jumped on a more internal orbit.
Now the required energy is equal to 9,
and the unnecessary amount
has been yielded in the form of a photon

*Figure 34. Quantum leap. An electron passing through a more
internal orbit loses energy, and yields it in the form of a photon.*

166

Between "Twin" Photons the Quantum Interweaving Is Established, That Is the Entanglement

A very long diatribe was grafted between the supporters of the special relationships between the photons emitted by a quantum leap, and the detractors. Among the latter there was also Einstein who, together with two other scientists (Podolski and Rosen) had devised, in 1935, a thought experiment to demonstrate the impossibility of the theories predicted by quantum mechanics.

Einstein defined predictions of particle behaviour at the quantum level as *"spooky action at a distance"* and denied them with determination. A thought experiment consists of an experiment that is not intended to be carried out in practice, but is only imagined: its results are not, therefore, measured. Thought experiments are used when there is no technical means to carry them out in practice or when there is no economic means available.

Almost thirty years later, in an article by John Stewart Bell, published in 1964 with the title *"On The Einstein-Podolsky-Rosen Paradox"* the author, supporting Einstein's ideas, emphasized that the non-local character of quantum mechanics, i. e. the phenomenon of entanglement, was incompatible with physical reality and concluded that, if this *seemed* to happen, then the studies were incomplete, and certainly there were no elements of calculation, because it was impossible for entanglement to possess the characteristics that were attributed to it.

However, Bell's article left open a possibility, because it suggested a new experiment, a simplified version of the EPR experiment, which could be realized by measurements of photon polarization.

Maybe he did not expect it, but someone took him at his word, and the suggested experiments took place. Since the 1980s, many have been carried out, and all provide experimental evidence that, at a quantum level, the localism of classical physics is no longer valid and non-local properties take over. The predictions regarding the characteristics of the entanglement have been fulfilled.

17. The Alain Aspect Experiment

The first to take up Bell's challenge and set up the experiment that confirmed the non-locality of the entangled particles was Alain Aspect.

Aspect knew that such experiments had already been drafted in the USA, between 1972 and 1976. The most recent had been performed by Fry and Thompson and had given confirmation results for the non-locality of the quantum level.

Aspect set his experiment according to the indications proposed by Bell, who suggested how to do it (it was an improved version of the EPR experiment) but remained skeptical about the results.

Things did not go as Einstein and Bell thought. In 1982 Aspect published its results, which definitively established how quantum reality was governed by non-locality.

Aspect's experiment was to excite an atom so that it could produce two photons simultaneously, headed in different directions. He used a bundle of calcium atoms as photon sources. Atoms were excited by a laser, and this forced an electron of each atom to skip two energy levels (two orbits). When the electron lost the energy no longer needed, because it approached the nucleus by two levels, it emitted a pair of photons (two light quanta).

These photons born from the same quantum leap, that is from the same electron, are called *correlated photons*. We can consider them as twins.

These two twin photons behave according to the characteristics of the non-locality we talked about in the previous chapters, confirming the non-locality of the quantum level. Non-locality means that at that level neither time nor space exist anymore: energy and information do not travel at a predetermined speed (at most the speed of light); indeed, they do not travel at all: they are both here and there, at the same time. Moreover, they are there in an eternal now, because in the non-local level there are no before and after. Indeed, if we want to be precise, there is not even the now.

These amazing properties make us understand why the theory of entanglement had to face so many difficulties to assert itself.

Today, about thirty years later, Alain Aspect's experiment has been repeated in countless ways and on countless occasions. Entanglement has been created not only between two, three or more particles, but recently between millions of particles, as reported in October 2017 in an article published on the Meteoweb website, signed by Filomena Fotia, entitled *"Physics: Forerunner Experiment for Quantum Communications, "Entangled" the Behavior of 16 Million Atoms"*.

The Principle Behind Entanglement

Observing two photons created together, that is related, we note that, as if they were human twins, they keep a mysterious bond between them. But not quite like two human twins: much more. Even separating them, each of the two remains connected to the other.

According to the rules of physics, each photon can have a value, called *spin*, which mathematically is equal to *half* and can be positive ($+1/2$) or negative ($-1/2$). These are opposite values. To better understand, we will give examples by imagining that the two photons are called Daniel and Antonia, that is, they are one male and the other female.

Once generated, the two photons leave in two opposite directions, each one on its own account: for now, the only link we attribute to them is that of being correlated, i. e. born from the same father electron (or mother, if you prefer).

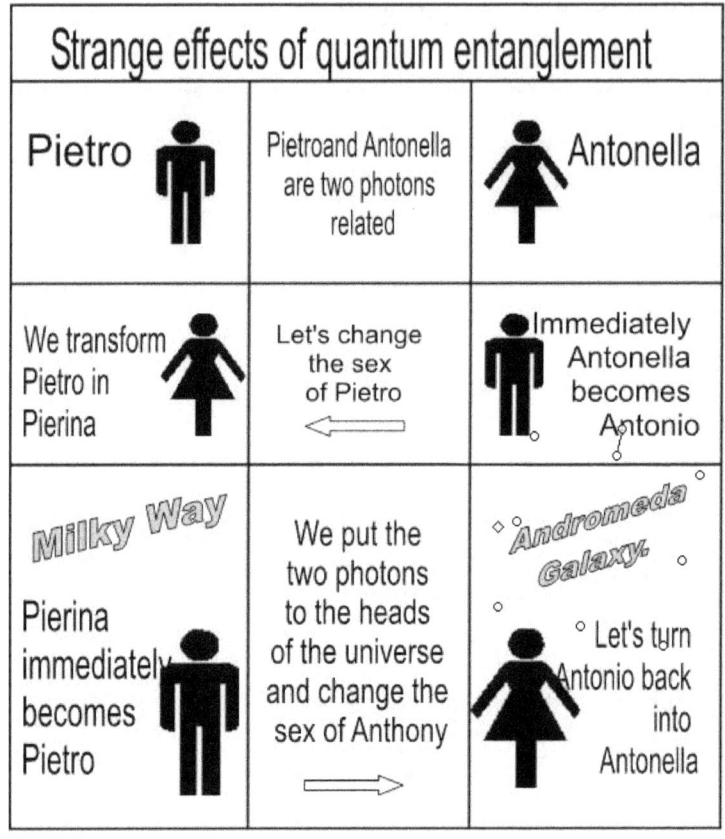

Figure 35. Two correlated photons, even if divided by immense distances, continue to behave as if they were one. This confirms the non-locality of entangled reality, not subject to the rules of time and space.

What happens if we intervene on Daniel making him change sex, so that he becomes Danielle?

Immediately even Antonia changes sex and becomes Antony!

You may think: Well, Antonia was very close to Daniel, so she noticed it and changed sex too, in response to the shock.

But if we repeat the experiment keeping Danielle here and moving Antonio on the other side of the world, the same thing happens.

Then you will say: - But, being on the other side of the world, Antonio will take some time to notice, and will turn into Antonia only after a few moments.

Wrong. Antonio changes at the same time as Danielle, wherever he is in the universe *(see fig. 35)*.

Equivocal Habits of Entangled Photons

Now imagine that Daniel has to leave for a business trip, leaving Antonia alone. On his journey he meets Louise, who is part of another entangled couple with Michael. At that moment, Michael is also far away *(see fig. 36)*.

Maybe it is the liking, maybe it is the distance, Daniel and Louise become entangled between them.

Do you think that Antonia and Michael will be indignant with jealousy? Not at all. Immediately Antonia becomes entangled with Michael, even though she has never seen him in her life.

173

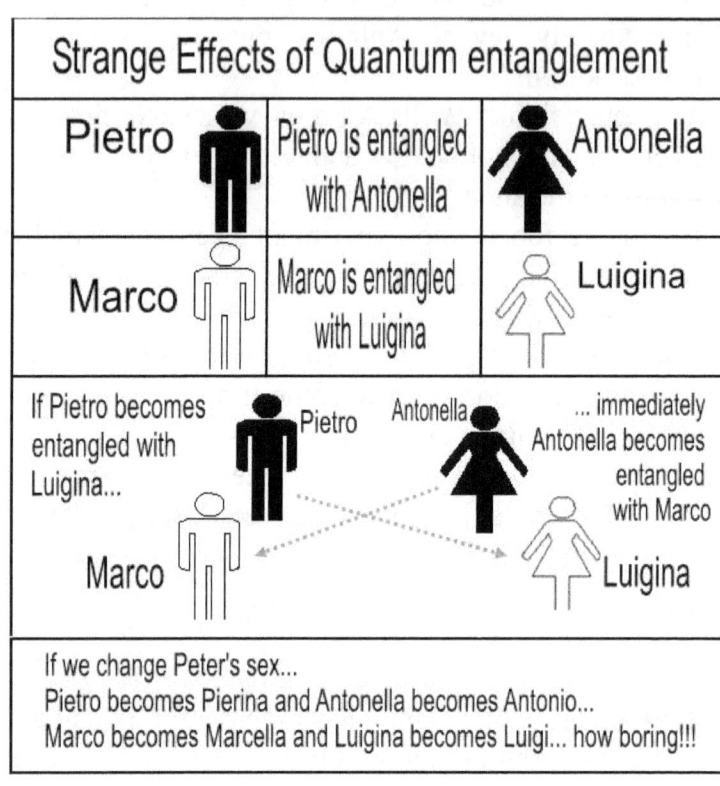

Figure 36. Weird effects of entanglements. Entanglement between more than two related photons.

So if, with the usual experiment, we change Antonia's sex to Antony, immediately Michael also becomes Michelle, Daniel becomes Danielle and Louise becomes Louis.

Complicated? Probably yes, but that is the way things between particles work.

Fortunately, you will say, it only happens between correlated particles. Sure, but in fact *all* the particles in the universe are correlated, since they were all born from the same Big Bang.

That is why in the non-local universe everything is a whole, nothing is independent. We are all connected to the whole universe: at the non-local level there are no knowledge or time barriers. All knowledge of the universe is simultaneously available in the non-local.

We are immersed in it and we are part of it. It is not uncommon for crumbs of this knowledge to slip from the non-local to our conscious mind. A question that many people ask is: do all these sex-changing particles cause no reaction in the universe?

Definitely not. What we have proposed, in an example, as a change of sex, is in fact the change of spin, i. e. the inversion of the direction of rotation. Whichever direction the particles spin, they do not produce sensitive effects on the functioning of the universe, as long as they spin. We have established some conventions, so things have to turn in a certain way, because we are used to seeing them turn like this. Do you think that if a cook turns mayonnaise from right to left or from left to right something changes in the final result?

If you suddenly change the clocks by establishing that the hands should turn counterclockwise, would something change in the course of hours?

18. The Implications of Entanglement

Entanglement is all here, but let's see what the consequences are. The first is that the most important beliefs of traditional physics are dismantled.

Entanglement Dismantles Mechanism

According to a property of classical physics, mechanism, for a piece of information to travel between two bodies of matter, a means by which it must be transported is necessary. In addition, its speed can never exceed the speed of light, which is 300,000 kilometers per second. But in the entanglement, there are no wires connecting Daniel and Antonia, and in any case the information is absolutely contemporary, i. e. it travels much faster than the light.

Entanglement Dismantles the Arrow of Time

Between an action and its response a fraction of time must pass, even infinitesimal, because first the facts happen and then the consequences occur. But

Antonia's answers to Daniel (or vice versa) do not require any time, they are contemporary.

There Is No Space or Time in Non-Locality

Therefore, elementary particles live in a dimension in which time and space do not exist: they are however connected (entangled) between them beyond space and time.

How is this possible? How do Antonia and Daniel notice the change of their twin, where does this knowledge come from and how does it form? One of the answers states that the two are only apparently separate, but continue to be part of a single reality. Or, even if separated, they are united by a single great universal Mind so they can communicate as if they were still united in the electron from which they were born: they do not have two separate minds, but they are part of the same mind. The non-local level is the great, unique mind of the universe.

The three levels of reality or cosmology of reality.		
Physical level or physical cosmos.	Quantum level or quantistic cosmos.	Non-local level or psychic cosmos

Level of things subject to the rules of time and space.	Middle Earth. Things subject to non-locality rules.	Level of pure psyche not subject to the rules of time and space.

Figure 37. The three levels of reality. With reference to the experimental results on entanglement, we can confirm with certainty the existence of three levels of reality of which, until a few decades ago, we knew only the first.

Part. VI. Extrasensory Perception

"Science cannot solve the ultimate mystery of nature. And that is because, in the last analysis, we ourselves are a part of the mystery that we are trying to solve."
(Max Planck, initiator of quantum physics)

Speaking of extrasensory perceptions, we can ask ourselves mainly three questions:
- Do they really exist?
- How do they occur?
- What exactly are they?

If we consider all that has been said so far, there is not much left to say about the first two questions. It is obvious that they really do exist, and they occur as episodes of communication between personal consciences and that world that we have defined as non-local level.

A small summary of the non-local characteristics is contained in figure 38. It is necessary, because the succession over the millennia of different cultures, although having always considered it existing, has identified it with a wide variety of names and characteristics. The most common have been mentioned in the previous chapters, but for the sake of clarity I have summarised them in the figure. This is to make it clear that often, even when we use different names, we are talking about a single reality.

In its contents, this figure does not want to be irreverent towards any faith, belief or spirituality. That is why I have divided the contents by gathering them under the heading "Secular Vision" or "Religious Vision". I think it is evident that the hypotheses of the secular vision are nothing but confirmations of those of the religious vision, and vice versa.

Without dwelling on the first two questions, we want to answer the third.

Non – Local Properties. *Psychic Cosmos*		
Names	**Contents**	**Interactions**
SECULAR VISION Place of ideas Anima Mundi Unus Mundus Collective consciousness Universal mind	**SECULAR VISION** Information Energy Ideas Archetypes Force fields Morphic fields Morphogenetic fields	**SECULAR VISION** It generates synchronicities (communication events between consciences)
RELIGIOUS VISION God Holy Spirit Tao Atman	**RELIGIOUS VISION** the Word the Verb Energy Wisdom The Creative Force	It responds to force field stresses It generates prophets and prophecies. It promotes knowledge

Figure 38. Summary of the non-local level properties. The term non-local means that it cannot be physically placed anywhere.

181

19. The Sixth and the Seventh Sense

We have always been told, since Aristotle's time, that the human body has five senses: sight, hearing, touch, taste, smell. These senses are those that allow us to physically interface with the surrounding world, to obtain useful information for survival.

In the popular wisdom, there is another sense, the sixth, which does not use physical mediators (hands, tongue, eyes, etc.) to do its job, but it does it equally well. This is *intuition*.

Intuition is that wealth of knowledge that we have accumulated in our life experience, and manages to suggest us quite often what we can trust and what not, what should be done and what not, ultimately how to risk as little as possible in the face of uncertain situations. The more experience we have, the more intuition works. Therefore intuition, if on one hand it can be considered extrasensory, because it is not mediated by any physical organ, on the other it is not for two good reasons. The first is that it does not introduce into our knowledge unknown and external cognitions, but more simply re-elaborates the information we already have; the second is that it is mediated by our brain, which has its physical consistency.

Thus, the sixth sense operates rationally on known data; on the contrary, the seventh sense elaborates in our consciousness, in a completely irrational and unpredictable way, data and information that have often never even touched our mind and our knowledge. That is, the seventh sense introduces us to realities that are completely alien to our daily lives and our experience.

Cosmos

The term *cosmos* in philosophy means an ordered or harmonic system. The origin of the word is the Greek *kósmos*, which means *order*. In the scientific language, cosmos is considered synonymous with the *universe*.

Cosmology studies the material structure and the laws governing the universe conceived as an ordered set. In the philosophical field it is interested in the universe in reference to space, time and matter.

Cosmology has its historical origins in religious narratives that deal with the origin of all things, and in the great pre-modern philosophical-scientific systems, such as the Ptolemaic system.

Modern cosmology, on the other hand, is the science that studies the universe as an ordered whole, its origin and its evolution. In this sense, it is closely connected with philosophical cosmology, but tends to correct the many metaphysical or religious theories about the origins of the world.

Cosmology is currently a physical science in which several disciplines converge, such as astronomy, astrophysics, particle physics and general relativity.

In this book we talk about the cosmos, seeing in this term the set of all realities, from the visible universe (physical cosmos) to that of the elementary particles (quantum cosmos) to that of the non-locality (psychic cosmos).

Definition of Extrasensory Perception

The term seventh sense can well define some properties more scientifically known as extrasensory perceptions, or ESP (*abbreviation of Extra Sensory Perception*). These are:

- **Precognition**, or the ability to predict the future.
- **Clairvoyance**, or the ability to visually perceive things that are not normally visible.
-**Telepathy**, or the ability to communicate with thought.

The study field of extrasensory perceptions is commonly called parapsychology.

These capabilities are often considered as "information flowing *from the outside to the mind*".

Another kind of ESP, telekinesis, is considered as "information flowing *from the mind outwards*".

In fact, telekinesis is the ability to manipulate matter with the mind, and is expressed through physical phenomena, from levitation to bilocation to the folding of metal objects, teleportation, time travel.

As far as precognition, clairvoyance and telepathy are concerned, many studies have been conducted, particularly in the century before the present, to verify their correspondence to reality. The most commonly used experimental verification method was (and is) the so-called *ganzfeld experiment*.

Ganzfeld Experiment

Ganzfeld is a German term meaning "*complete field*". This term was coined in 1930 by Wolfgang Metzer, a German psychologist and one of the founders of the Berlin Gestalt School.

It indicates an experimentally constructed perceptual field to create the optimal conditions for the study of extrasensory perceptions. In a typical ganzfeld experiment, the subject to be examined wears a set of headphones (through which white or pink noise – static - is played) and halved ping-pong balls over the eyes, having a red light shone on them. This is to ensure isolation from any possible external distraction.

After a certain period of time in this state, an attempt is made to transmit images to the subject through an extrasensory channel, for example by telepathic means.

While a person located elsewhere concentrates on the image to be transmitted, the receiver describes what he is seeing in his mind.

In many experiments, at the end of the transmission attempt, four images are presented to the subject, asking him to indicate the one closest to the image he thinks he has received *(see fig. 39)*.

In these cases, the statistical success rate would be 25% (one in four). If the receiver guesses the transmitted image, choosing it among the four presented to him, in 25% of the cases, the result is average and the experiment has failed.

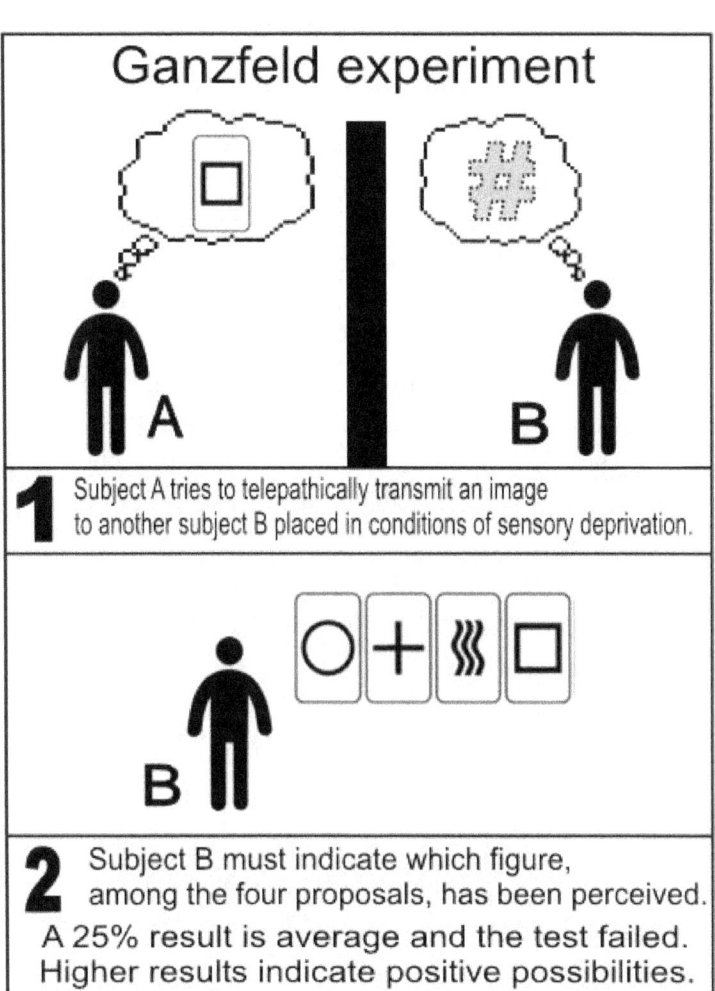

Figure 39. Scheme of a typical ganzfeld experiment.

CICAP (*Comitato Italiano per il Controllo delle Affermazioni sulle Pseudoscienze)* writes about these experiments on its website:

While a person located elsewhere concentrates on the image to be transmitted, the receiver describes what he is seeing in his mind.

In many experiments, at the end of the transmission attempt, four images are presented to the subject, asking him to indicate the one closest to the image he thinks he has received *(see fig. 39)*.

In these cases, the statistical success rate would be 25% (one in four). If the receiver guesses the transmitted image, choosing it among the four presented to him, in 25% of the cases, the result is average and the experiment has failed.

CICAP (Comitato Italiano per il Controllo delle Affermazioni sulle Pseudoscienze) writes about these experiments on its website:

"In the first ganzfeld experiments the success rate was 33-34%. We then found ourselves faced with a statistical anomaly which, according to some, proves the existence of non-normal channels of information communication"

In 1994 Bern and Honorton published an article entitled "Is there extrasensory perception? A repeatable proof of an anomalous information transfer process ".

These conclusions were later widely criticized by other scholars.

In fact, in all the experiments concerning the psychic field, there is always a problem of repeatability. This is because emotionality plays a decisive role, as we have already seen in chapter 9.

Any contact with the non-local is favored by particular stress conditions or critical moments in life. We know that telepathy phenomena are absolutely more frequent among people linked by strong bonds of friendship or kinship, reaching a peak between mothers and children. Instead, in the experiments, often unknown persons are asked to "coldly" share images normally devoid of affective meaning.

Another role is played by habit and inurement: people who constantly repeat the same experiment several times feel emotionally involved to an extent inversely proportional to the repetitions, that is, they end up surrendering to boredom.

20. Telepathy

Telepathy, also called *thought transfer*, is a form of non-verbal communication, that is conducted directly from one mind to another, without the use of other tools or devices that create a physical connection between the people involved. The term derives from two Greek words, *tele* which means distant and *patheia* which means feeling, perception.

In practice, telepathy is about the fact that physically separated minds can connect with each other beyond distance or circumstances. This ability to share emotions and sensations at a distance is strongly contested by materialistic science. In fact, telepathy contradicts the assumption that there is nothing outside the brain: thoughts, emotions, desires and all that is psychic is born and dies inside the skull box. There are no "thoughts" that can flutter from one brain to another. Every experience of telepathy is considered, in the most benevolent case, an illusion. However, often, witnessing episodes of telepathy, one is considered to be the victim of schizoid personality disorder.

Despite this, many people claim to have had telepathic experiences. In various surveys conducted in the US or the UK, 40 to 60% of respondents say they have had personal experiences of telepathic kind.

There seem to be two types of telepathy. A first type is the one that occurs between people who are physically close, who know each other and already interact with each other. It is the classic case in which one says to the other: "You read my mind", meaning that his intention was perfectly understood and deciphered.

The second case is referred to two people physically distant, when one feels the *"call"* of the other through a thought that binds them, his image, or even by hearing his voice.

This happens mostly between people who have close ties, when conditions of need or danger are manifesting.

Joseph Rhine Experiments

Classical experiments on telepathy are carried out through drawings. A person, placed in an isolated room, chooses an object and makes a sketch of it. Then he concentrates trying to send the produced figure to the receiving person, placed in another room: this second person tries to receive and replicate the transmitted figure. Finally, the images are compared to identify all the common aspects.

The first studies on telepathy were conducted by the British Society for Psychical Research towards the end of the 19th century.

In 1930, the first parapsychology laboratory was set up at Duke University in Durham, USA, where experiments using Zener cards were conducted by the famous Joseph Rhine.

Rhine was the first to apply experimental and statistical methods in the search for the existence of a mind-to-mind transmission or telepathy, which he called extrasensory perception. He is certainly one of the pioneers of parapsychology.

Zener cards are a special type of card, designed by Karl Zener, a psychologist, to be used by Joseph Rhine in his experiments.

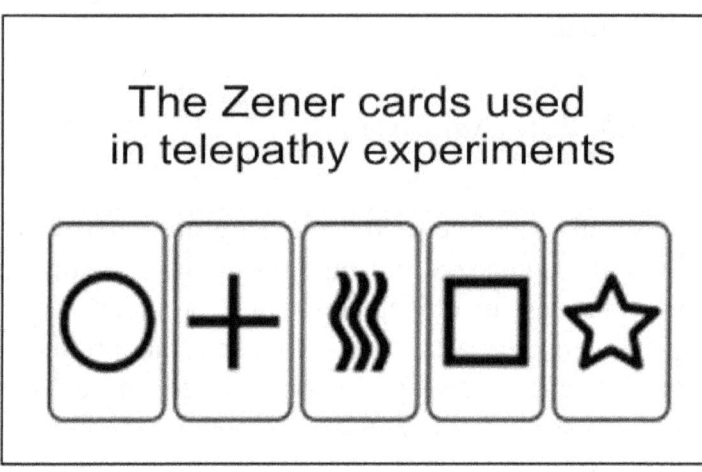

Figure 40. Zener cards, used in many parapsychological experiments. This is a series of five symbols repeated five times, for a total of 25 cards.

The deck is made up of 25 rectangular cards as the common playing cards. They are divided into five groups bearing a sign: circle, cross, square, star and waves. The deck includes 5 equal groups (see fig. 40).

The experiment conducted by Rhine was based on two people. One was asking the other to guess the card he was picking from the deck.

In the precognition experiments neither of them knew the card, which remained turned on its back.

Instead, in telepathy experiments, the first subject saw which card it was and tried to mentally transmit it to the other subject.

Clearly, the two people were placed in different rooms.

According to a statistical hypothesis, and according to the law of large numbers, the probability of guessing a card of the deck must be around 20% (1 out of 5).

However, Rhine achieved significantly more than this percentage. The results of his research were published in 1940 in a book known as ESP-60, which was adopted as a textbook by the University of Harward for psychology courses.

Upton Sinclair Experiments

Upton Beall Sinclair was born in Baltimore in 1878. Living among constant economic hardships, he was driven since his youth towards socialist ideals. He achieved a resounding success with the novel

Figure 41. Upton Sinclair and the cover of his book Mental Radio, published in 1930 with a preface by Albert Einstein.

The Jungle (1906), about the scandalous workers condition in Chicago's livestock markets. The book, that was defined by Jack London "*the Uncle Tom's Cabin of wage slavery*", was reviewed by Churchill and received words of admiration from Bernard Shaw.

In 1930, together with his second wife Mary Craig Sinclair, he carried out parapsychological experiments concerning the duplication of images through telepathic communication.

In his book *Mental Radio* (*see fig. 41*) he describes the experiments made. Out of 290 attempts to duplicate the images, his wife Mary successfully reproduced 65 of them, achieved 155 "partial successes" and 70 failures.

His book was a great success. The German edition included a preface written by Albert Einstein, which read as follows:

In 1930, together with his second wife Mary Craig Sinclair, he carried out parapsychological experiments concerning the duplication of images through telepathic communication.

In his book *Mental Radio* (*see fig. 41*) he describes the experiments made. Out of 290 attempts to duplicate the images, his wife Mary successfully reproduced 65 of them, achieved 155 "partial successes" and 70 failures. His book was a great success. The German edition included a preface written by Albert Einstein, which read as follows:

"I have read the book of Upton Sinclair with great interest and am convinced that the same deserves the most earnest consideration, not only of the laity, but also of the psychologists by profession. The results of the telepathic experiments carefully and plainly set forth in this book stand surely far beyond those which a nature investigator holds to be thinkable. On the other hand, it is out of the question in the case of so conscientious an observer and writer as Upton Sinclair that he is carrying on a conscious deception of the reading world; his good faith and dependability are not to be doubted. So, if somehow the facts here set forth rest not upon telepathy, but upon some unconscious hypnotic influence from person to person, this also would be of high psychological interest. In no case should the psychologically interested circles pass over this book heedlessly".

These few words of Einstein are a very clear example of what can distinguish a scientist with a luminous mind from the mass of mediocre minds. Einstein was always animated by deep curiosity and thirst for knowledge. He did not exclude anything a priori, but in any case, he did not renounce to investigate, to experiment. This led him to elaborate theories, such as that of relativity, which no one had imagined and in which no one would have believed except one like him, a mind free from prejudices.

Figure 42. René Warcollier, French engineer and parapsychologist, and the cover of his successful book Mind to Mind, on his telepathic experiments, published in 1948, but still reprinted today.

With the same criterion Einstein, while not sharing the implications of quantum physics, did not reject it a priori but conceived the EPR experiment to verify how real there was in it even beyond any credible hypothesis.

René Warcollier Experiments

Warcollier was born on 8 April 1881 in Omonville-la-Rouge, Paris. He graduated in chemical engineering in 1903 at the Ecole Nationale Supérieure de Chimie. He patented several processes related to the synthetic production of precious stones and created special screens for cinema projection.

He took an active interest in parapsychology, so much so as to have been president of the Institut Métapsychique International.

His main telepathy studies were conducted in such a way that one or more "senders" observed an image, while one or more "receivers" tried to reproduce it. Much of his work after 1922 included many senders and receivers spread throughout France; he also used senders and receivers located in different countries, such as France and the United States, or France and Britain.

His experimental method was a step forward compared to the previous ones, because he used drawings of real objects. So, Warcollier discarded the statistical method and attempted to obtain direct evidence for telepathy. He claimed that, according to his experiments, men convey thoughts better than women, while women are better recipients. He also stated that younger people are more sensitive to mental impressions than older people.

His experiments focused on how the original images were perceived, distorted or otherwise deformted by the receivers. He realized that they were not transmitted as whole photographs, but the various components of the figure were broken up and shuffled. This is confirmed by modern neuroscience, about the way in which visual images are perceived and reconstructed by the brain. According to Warcollier, telepathic images emerge from the unconscious to the consciousness, and here they are processed as dream pictures. He, therefore, concluded that images transmitted telepathically are distorted and less clear than those normally perceived by sight.

Warcollier reported his experiments in numerous publications, mainly in 56 articles in the Revue Métapsychique. They were also published in his many books, including *La Télepathie* (1921) and *La Métapsychique* (1940, 1946) and especially *Mind to Mind*, published in 1948, but still reprinted today *(see fig. 42)*.

Entanglement and Telepathy

All the experiments we have reviewed (only some of the best known) were conducted by scholars who did not yet know the entanglement. This novelty makes possible all the assumptions about a non-local reality, that is, about a psychic cosmos in which ideas and thoughts can circulate freely, even in the form of telepathic transmissions.

We can imagine a force field, linked to each person, that connects his physical reality, that is his mind and his conscience, to the psychic cosmos, where he can come in contact with all the force fields of the universe.

The more two people are connected to each other, the more their force fields establish understandings on our physical level and continue these understandings on a non-local level, i. e. in the psychic cosmos. In all the experiments, it has always been found that the phenomenon of telepathy is much stronger, the more there is a bond of friendship or kinship between the two people, with a maximum between mothers and children.

In figure 43, an elderly mother is experiencing a difficult time due to an accident or serious illness. Her personal force field, loaded with anxiety, is excited and transmits excitement to her child's personal field

A suffering field is established between the two fields. The son perceives the serious situation as a "call" that can be explained in different ways: he hears his mother's voice, he sees her, he sees the dangerous situation in the details, he sees a symbolic image that makes him imagine the danger.

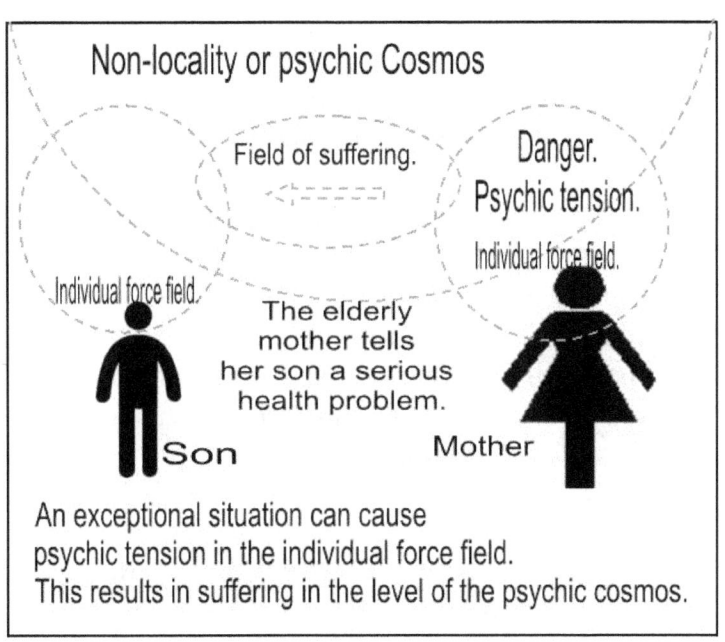

Figure 43. Example of a telepathic episode. The mother's personal force field is in a state of tension due to a situation of possible danger. A suffering is created with her son's personal field, who "receives a call".

The mother-son situation is also very common among twins. There are many cases of twins who, although separated after birth, have recorded incredible behavioral similarities in their lives. Many believe that these are curious but possible coincidences.

I can cite, as an example, the case of the Ohio twins (USA), Jim Lewis of Dayton and Jim Springer of Lima, who were separated at birth and did not know each other. The list of similarities is indeed very rich. It is reported by the two scholars who were interested in the case, Colin Wilson and Damon Wilson in *The Encyclopedia of Unsolved Mysteries*.

- They both were named Jim by their adoptive parents.
- Both had the habit of biting their nails and suffered or had suffered from migraines.
- They both smoked the same brand of cigarette.
- They both had the same passion for DIY.
- They both spent their holidays in the same area.
- They both called their kids James Allan.
- They both had a dog named Toy.

- Both had worked at McDonald's, at a gas station (not of the same chain) and as sheriff's assistants.
- Both had divorced and remarried.
- Their first and second wives had the same names (Linda and Betty).

My opinion is that in their lives these twins continued, perhaps unconsciously, to maintain telepathic relationships, which led them to adopt extraordinarily similar life choices.

Sensory Coincidences Between Twins

The CICAP website contains examples of sensory coincidences between twins, calling them "decidedly sensational".

"These are cases, also decidedly sensational, presented both by pairs of twins MZ (*monozygotic*) still joined, and by twins always MZ naturally and voluntarily separated in adulthood. They usually take the form of unusual coincidences in the perception of sensations that seem to derive from experiences that only one of the twins is actually experiencing, but that the other also experiences without apparent reason. This is the case, for example, of the labor pains that appeared prematurely in one of the twins and were also perceived by the other (not pregnant) who was miles away and completely unaware of the event. Like the case of Thelma Furness, who had to face a premature birth while she was in Europe, while her twin in New York Gloria Vanderbilt, totally unaware of the emergency childbirth, felt severe abdominal pains. Or the sudden fall of a twin in conjunction with a similar accident that happened to the other that was several kilometers away. Like the case of Ross Mc Whirter, shot in the head in London on the evening of November 27, 1975, while his brother Norris, an identical twin who was 30 miles away, suffered severe headache at the same time".

(https://www.cicap.org/n/articolo.php?id=273421)

21. Personal Force Field

The concept of personal force field, mentioned in Figure 43, deserves some further investigation.

A force field consists of energy and information. Consequently, a person's force field contains his energy and all the information related to his Ego and his Self.

This information is his history, his experiences, what he has been, what he is and what he will be. Energy is the component that makes this information alive and operative. The energy and information of a living being could not exist, if this person did not have a physical body.

In fact, the information is given by the entanglement that connects and intertwines in a large unit all the particles that are part of his current physical body, but also those that were part of it in the past.

We have seen how two photons "born together" remain bound by a non-physical but psychic bond, which determines such behavior, as if they were one thing, even if separated. This link does not only apply to two particles, but to all the particles that somehow were born together, or otherwise shared their birth and existence closely with others.

Thus, we have observed that all the matter of the universe, composed of particles born from the initial explosion of energy in the Big Bang, are connected to each other so as to constitute in the non-local dimension a *Anima mundi (Soul of the World)*, a great universal mind composed of all the information and all the energy of the universe.

At a lower level, all the particles that have been, are now or will be part of a living being in the future, constitute all its information and all its energy, its personal *Anima mundi*. Maybe, just his soul.

In the figure 44 the *Anima mundi* of the single individual, which we can simply call soul by similitude, is absolutely personal, because it contains all its information and its history, contained in the matter that composed its body and bound by an indissoluble entanglement.

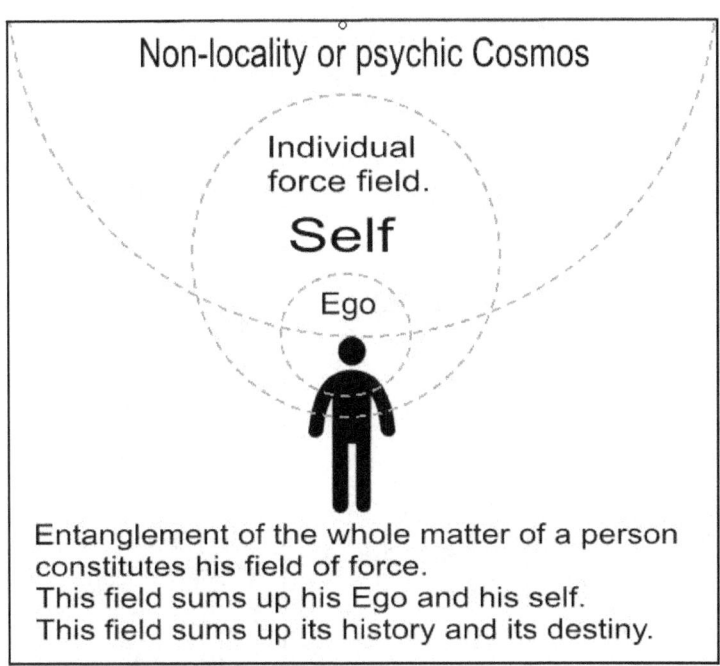

Entanglement of the whole matter of a person
constitutes his field of force.
This field sums up his Ego and his self.
This field sums up its history and its destiny.

Figure 44. The force field of each element of creation composed of matter is part of the psychic cosmos.

Each creature interacts with many force fields.		
Morphogenetic field.	Morfic field	Individual force field.

Morphogenetic field leads the physical development in the various moments of growth.	Morfic field assures each species the correspondence to the psychological characters of the species.	Individual force field contains the personal story written by the individual himself.

Figure 45. The main force fields that accompany creatures in their existence.

By the term creature, we mean here every aggregation of matter endowed with its own identity, be it mineral, vegetable or animal. A river, a daisy or a grasshopper, as well as a man, exists in the three levels already indicated: physical, quantum and psychic, and participates at each level through what we have generically called force fields, composed of energy and information.

In Figure 45, I would like to summarize the various types of fields in which each creature participates. Each force field can be better distinguished as morphogenetic field, morphic field or personal force field, it being understood that there can be many other types of field, some known by different names, depending on the authors who proposed them, others still unknown.

The morphogenetic field contains the constructive project of the creature, that is the procedure according to which the various chemical elements that compose it are arranged to allow its development, according to the characteristics of its species. This field interacts with creatures as they develop, and can remain affected by it by passing on updated information to subsequent creatures. That explains the mutations.

The morphic field contains the psychic characteristics of the species: instincts, tendencies, cultural and spiritual gifts. This field, which is different in species and variety, also interacts with individual creatures. By increasing the cultural evolution of the species, this increase is transmitted to the subsequent elements of the same species. Morphic fields can be considered similar to archetypes, since they also exist before individual creatures, and they too vary as the species evolves.

The persoal force field contains the whole story of the creature, different from that of all other creatures of the same species. It is a story built by living day by day, according to variations due to external events or, in the case of human beings, to free will. The personal force field is closely linked to matter, that is to say, to the body of the creature. In fact, many experiences are of a physical nature, that is, they are due to the five senses. As far as man is concerned, it is evident that, beyond the physiological processes carried out mostly unconsciously, most actions derive from personal choice, and this confirms that the personal force fields are different for each one.

Because of these characteristics, the concept of "personal force field" can be compared to that of "soul".

And What Happens When the Creature Dies?

The question we can ask ourselves is: what happens to the personal force field, when the creature dies?

Absolutely nothing. In fact, death does not involve the annihilation of matter, but only its disintegration.

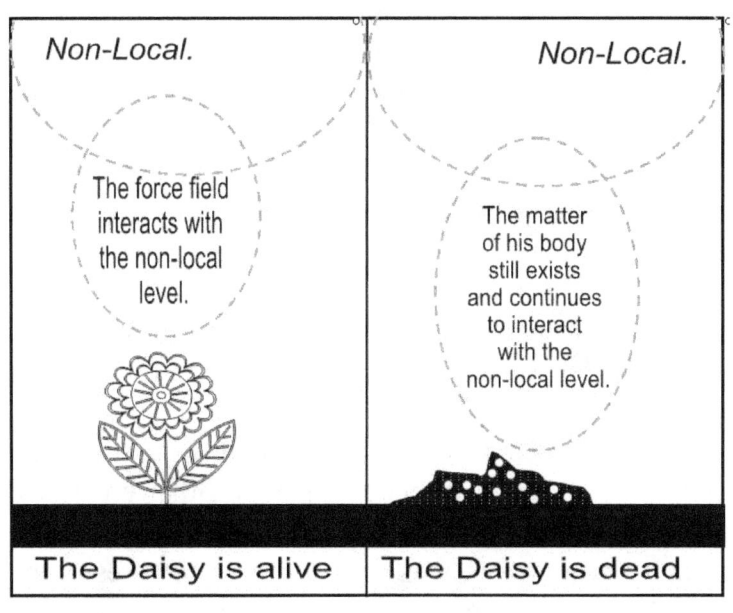

Figure 46. The creature force field does not cease to exist, as long as the matter that constituted its body exists.

Today, cosmologists like to say that we are made of star dust. Actually, it is. The atoms of our body may become part of other bodies, but the entanglement that united them will not cease to exist, just as the force field does not cease to exist.

Simply, it remains in non-locality, in the psychic cosmos, it contributes to the enrichment of the ideas that are contained in it and carries with it forever the history of what has been. Prodigiously, each atom, each particle can participate in multiple force fields, when it is reused in the construction of other creatures.

If it is true, that after the death of a creature, the matter of its body can become part of other creatures, it is also true that the same matter, that is part of the creature now living, was part of other creatures in the past. No one has the privilege of "being brand new": we are all built by recycling material, already used millions or billions of times.

Moreover, it is easy to understand how, already in the course of our lives, an infinity of waste products of our body, such as cut nails, sweat, the aerosols of our sneezes, go to be part of other bodies, but continue to share and affect our force field. But this does not surprise us, in the universe where everything is one.

So, is a creature's personal force-field immortal?

Maybe. We know that matter is energy, and all the matter in the universe comes from an initial energy charge, exploded in the Big Bang.

It is possible that in an unimaginable time all the matter in the universe will again collapse into energy. When this happens, there will be no more matter. Probably, the information of every creature will disappear, but not the energy that will return to join the great, unknown Energy from which it all began.

What about good and evil? Perhaps, the force fields of the living creatures will carry in the Energy their imprint of good and evil. This will still create tension in the Energy, which will explode in a new Big Bang generating a new universe, in a cycle that will end only when the Energy is all purified and pacified. But then it will really be the end.

22. Prescience

Prescience is the ability to have knowledge of the future. Those who possess this faculty are able to acquire knowledge of events before they happen, and of places, objects or people placed in the future.

Under this heading we can group many, which are often used as an alternative to each other, but have different meanings.

Presentiment is a feeling that something will or is about to happen, even if we often do not know what.

Precognition (*a priori knowledge*) is the knowledge of a future event or situation. In a sense, precognition is a more evolved stage than mere presentiment.

The **Monition** is the knowledge of a phenomenon that is occurring at a distance at the same time as the monition is perceived.

Premonition is a strong feeling that something is about to happen, especially something unpleasant, about which we are warned. Often it is a feeling that accompanies a certain action that is being carried out or wants to be carried out, and lets us glimpse with certainty the consequences.

Clairvoyance (to see clearly) consists in the mental vision of facts that will occur in the future, as they will take place, almost like in a photo or a film.

23. A Tragic Sea Adventure

In 1827, the whaler Grampus sailed from Nantucket, an island in the United States of America, 48 km south of Cape Cod, in the State of Massachusetts. Captain Barnard knew that a stowaway, the young Arthur Gordon, a dear friend of his son Augustus, was hiding in the hold. Besides, even the captain was a friend of Arthur's father, for this reason he had accepted the deception. The agreement was that, once on the high seas, the presence of the young man would be "discovered", but he would remain on board, since he could no longer disembark. With this the young Arthur wanted to satisfy his spirit of adventure together with Augustus.

Everything went smoothly for a few days until Arthur, no longer receiving news or food from his friend in his hiding place, discovered that there had been a mutiny on board with truculent stories of murders, such as the case of sailors killed by ax strokes by the cook, deployed with the rebels.

Arthur, his friend Augustus and a repentant mutineer managed to thwart the revolt. Unfortunately, a storm severely damaged the ship, leaving it without government. The three friends, along with another young man named Richard Parker, had to leave it and find refuge on another ship that appeared by chance, which, however, turned out to be abandoned.

Without water and food, exhausted, the four decided to sacrifice one of them, so that the others could feed on it to survive.

A True Story. The Mignonette Shipwreck

In 1883 a wealthy Australian lawyer, John Henry Want, who was visiting England, had the opportunity to visit a small 16-metre yacht, the Mignonette, which was for sale. He was fascinated and decided to buy it. The problem was to get it to Australia, being a small boat and therefore unsuitable for that crossing. It was decided that the yacht would set sail on 19 May 1884 from the English port of Southampton to Sydney. The crew consisted of four people: Captain Tom Dudley, Edwin Stephens, Edmund Brooks and *Richard Parker*, a 17 year old boy with no sailing experience.

Figure 47. The case of cannibalism narrated in "The Narrative of Arthur Gordon Pym of Nantucket" in an illustration of the time.

THE NARRATIVE
OF
ARTHUR GORDON PYM
OF NANTUCKET

EDGAR ALLAN POE

Figure 48. Title page of an English edition of the story by Edgard Allan Poe.

When the Mignonette was about 1600 miles northwest of the Cape of Good Hope, an abnormal wave hit it and turned it upside down. It sank shortly afterwards, but the crew managed to escape on a lifeboat.

Unfortunately, because of the haste, they had brought with them nothing useful for survival. Left to their own devices in the middle of the ocean, they decided to sacrifice one of them for the survival of others. They did not draw lots. Captain Dudley and sailor Stephens decided it was up to young Parker, although the other sailor, Brooks, abstained. The reason for the choice was that Richard Parker was sickly and did not have a family to return to.

The three of them were saved thanks to the (involuntary) sacrifice of the young hub. They were later rescued by a passing ship, the Montezuma, off Rio de Janeiro, between 26 and 27 July 1884.

Back home, they were prosecuted.

On 22 December 1884 Dudley and Stephens were sentenced to death. Later, however, following a general uprising of public opinion in favour of the survivors, the sentence was reduced to six months in prison.

Figure 49. The yacht Mignonette, a small boat of 16 meters, protagonist of a terrible story of shipwreck, in an illustration of the time.

24. Literary Foresight

Thus, Edgard Allan Poe in his fantastic tale about the adventures of Arthur Gordon Pym, written in 1837, tells a case of cannibalism in which, following a shipwreck, a young hub named Gordon Parker is involved, eaten by the other three survivors.

Incredibly in 1884, 47 years later, the case really happens and a young hub named Richard Parker, escaped with three others to the shipwreck of the Mignonette, becomes food for the other survivors.

L'Enciclopedia Treccani definisce così gli episodi di preveggenza: "La premonizione è un'informazione paranormale relativa al verificarsi di eventi futuri. Più genericamente è un presentimento, un presagio, un segno premonitore".

The Treccani Enciclopedia thus defines the episodes of foresight: "The premonition is paranormal information concerning the occurrence of future events. More generically it is a presentiment, an omen, a premonitory sign".

When we feel that a certain event is going to happen, and then it really does happen, we are disconcerted. How did we manage to predict so accurately a fact, that would only happen in the future?

In fact, premonition or foresight is a mental dynamism that focuses on an event already contained in the non-local dimension. Many events that have happened in the past, or that will happen in the future, could suddenly appear very clear in every detail. In these cases, our consciousness has freed itself from the rules of time and space in force in our sensitive universe, to draw on the universal Mind which, being placed in the non-local dimension, knows no physical limitations.

In this context, cultural foresight, typical of those used to working with the brain as writers, creatives, but also scholars, is more easily able to penetrate the universal knowledge of the non-local by drawing information that, in our level subjected to space-time, may seem to anticipate future events. In addition to the case of Richard Parker's tragedy, there are many other cases of cultural foresight by writers. One of the best known is the one related to the sinking of the Titanic.

Futility. The Wreck of the Titan

Morgan Robertson is an American writer, born in Oswego in 1831, known for his stories based on sea adventures; certainly he became fond of this genre for its origins, in fact he was the son of Andrew, ship captain. It seems that Morgan was the inventor of the periscope, or collaborated in its realization. His celebrity, however, derives almost entirely from a story he wrote in 1898 under the title *Futility*. It tells the story of a large transatlantic liner that, due to the collision with a large iceberg, sinks in the Atlantic Ocean.

Figure 50. The shipwreck of the Titanic in a New York Times headline. On the right, the cover of an old Italian version of the

story by Morgan Robertson, Futility, renamed "The Wreck of the Titan".

Affinity between the (fantastic) shipwreck of the Titan and the (real) shipwreck of the Titanic, which occurred 14 years later	
Name	One ship was called Titan, the other Titanic.
Victims	Both shipwrecks cause thousands of deaths.
Cause of shipwreck	Both ships sink due to a collision with an iceberg.
Impact point	Both hit the iceberg on the starboard side.
Shipwreck site	In both cases the shipwreck takes place about 400 miles from Newfoundland.
Speed at impact	Both were traveling at similar speeds (the Titan 25 knots, the Titanic 22.5 knots).
Nickname	Both ships were defined as "unsinkable".
Propulsion	Both had three propellers and two trees.
Dimensions	The Titan was 244 meters long, the Titanic a little longer, 269.
Lifeboats	Both had not enough lifeboats for all travellers.
Departure	Both left in April.
Route	Both were on the route from the UK to New York.

The plot of the story is centered on the protagonist John Rowland, former US Navy officer, alcoholic and degraded, who works as a sailor on the Titan. This ship hit one night an iceberg and sinks, but John manages to save himself right on the iceberg with a girl. As expected, with her help, he will save himself, solve his alcohol problems and live the rest of his life happily.

What interests us, however, in this story, are not the personal stories of John Rowland, but the extraordinary coincidences that link the imaginary shipwreck of the Titan with the real one of the Titanic, which took place in 1912, that is only 14 years later.

Let's try to list them all.

It must be said that, following the actual shipwreck, the story was republished with changes to make it more similar to reality, but the similarities that I report here are related to the first version.

Extraordinarily, CICAP goes so far as to settle the matter by saying that it was quite simple for the author to imagine the story of the shipwreck of a large ship, since there were so many of them in circulation. No comment.

25. Presentiments

Each of us, at some point in his life, had some presentiment. Presentiments are often accompanied by physical sensations such as chills, goosebumps, but also by "butterflies in the stomach".

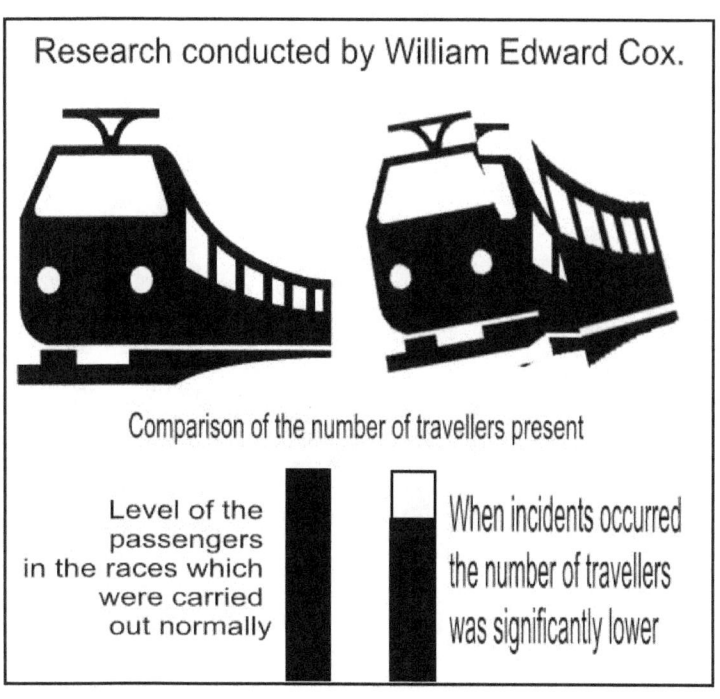

Figure 51 - According to a study by William Edward Cox, who carried out a comparison between travelers on trains in

different journeys, in cases where accidents occur, the number of travelers was significantly lower.

A researcher at Columbia University in New York argues that, indeed, the stomach hosts a large network of nerve cells, perpetually connected to the emotional part of the brain. It is obvious that an emotional force field, generated by a presentiment, also stimulates the stomach, so as to make the emotion perceptible.

Presentiments are very frequent and often save people's lives.

In 1956 the American engineer William Edward Cox published, in the number 50 of the *Journal of the American Society for Psychical Research*, an article entitled "Precognition. An analysis", containing the report of his research on precognition.

In practice, Cox had conducted a study comparing the number of passengers on trains that had made normal runs with that of passengers on 28 trains that had had accidents. The comparison showed a significant reduction in the number of passengers present on the routes affected by accidents, compared to those present on the same trains, that had normally travelled the route *(see fig. 51)*.

Cox had published dozens of articles and research on parapsychology, and had worked in the Parapsychology Laboratory at Duke University in Durham, North Carolina, so he was not an improviser, but an experienced experimenter. He passed away in 1994.

Some scholars say that presentiments are nothing but a manifestation of intuition. As already mentioned, intuition is not an extrasensory perception, because it draws on the knowledge already possessed, elaborating them to obtain indications of behavior. There is definitely a causal link between intuitive reactions and knowledge already possessed. In the presentiment, however, there is no causal relationship: even if experience tells us that sooner or later a train will have an accident, it cannot tell us that that very train, and that very day, will have an accident.

As is evident from the study by engineer Cox, passengers decreased just in coincidence of the days and the rides that suffered accidents. Therefore, not intuition but presentiment.

Dean Radin's Experiments

Dean Radin is a psychologist, director of the Consciousness Research Laboratory at the University of Nevada, Las Vegas.

He belongs to the group of parapsychologists with a solid scientific background who work in prestigious facilities, such as universities or other research centres.

In the mid-1990s Radin, together with his collaborators, began a series of experiments aimed at verifying the presentiment.

In the typical experiment, the emotional excitement of the subjects undergoing the test was measured, detecting with special instruments the variations of the skin resistance through electrodes connected to

232

the fingers, more or less as it is done in the lie detector's test.

Experiments by Dean Radin

To Volunteers Images from the pacific or shocking content are shown.

Quiet pictures.
The reaction
occurs 2-3 seconds
AFTER
viewing the image.

Highly emotional
images.
The test reaction is
two or three
milliseconds
BEFORE
the image is vision.

Figure 52. In Dean Radin's experiments, subjects controlled by electronic equipment could perceive the arrival of shocking images before they appeared on the screen.

The basic principle is that when a person changes his emotional state, the activity of his sweat glands and consequently the electrical activity of the skin also varies.

During the test, subjects underwent "emotional stresses" that were significant enough to produce significant changes in the sweat glands. To achieve this, they were subjected to small electric shocks, to sudden and unpleasant smells or, more often, they were shown images alternating photos of quiet landscapes or smiling children to photos of corpses torn apart during an autopsy, or road accidents, or with highly pornographic content.

Of course, the images suddenly appeared on a screen *(see fig. 52)*.

During the various sessions of the experiment, Radin and his collaborators noted that when subjects were shown peaceful images, they maintained their tranquility.

When, on the other hand, images with a high emotional charge were shown, the subjects reacted strongly, as could be detected by the connected instruments.

So far so good, where is the surprise?

It is in the fact that, measuring in milliseconds the time of their reaction to the shocking images, it was discovered that their emotional state varied a few milliseconds BEFORE the image appeared, just as if their unconscious knew that the incoming image was highly emotional.

Disaster Forecasting

The testimonies concerning the forecast of disasters are innumerable. We can recall one case above all, the landslide that submerged the mining village of Aberfan, in Wales, on 21 October 1966. The landslide of more than 150,000 cubic meters of coal mining slag, caused by the heavy rains of the previous days, submerged the town at 9:15 in the morning. The mountain of debris also hit the school where all the pupils were doing their classes. If it had happened an hour earlier, the school would have been empty. There were 128 children among the 144 victims.

In his book *The Extended Mind* the author Rupert Sheldrake writes as follows:

> "The psychiatrist J.C. Baker, who worked in the village after the tragedy, discovered that many people had had premonitions.
>
> He investigated 76 cases. Of these, 3 were about dreams. Some dreams were so vivid and horrifying that those who did them would wake up with a start shouting.
>
> Some people felt intense anguish. A ten-year-old girl, Eryl Jones, killed in the tragedy, told her mother two weeks earlier: "Mom, I'm afraid of dying" and added that she would die with her friends.

Figure 53. The first aids in the village of Aberfan (from the Londranews.com site)

The mother wondered why she spoke that way. Then, the day before the tragedy, Eryl said to his mother: "I dreamed that something black had gone down and crushed the school". At the request of her mother, Eryl was buried next to her friends. "

26. Other Types of Prescience

There are also other types of prescience, not linked to tragic events. For example, it is very common to think of someone we had practically forgotten, or maybe recognize him in an old photo, and shortly thereafter, the same day or the next day, meet him or receive a phone call, an e-mail or a letter.

There are people who always know in advance when someone is about to arrive, especially if there are bonds of affection or kinship between the two.

Apparently, presentiments are also experienced by animals. It seems that many animals can predict the arrival of earthquakes, although we do not know if this really depends on prescience or on particular sensitivities that make them able to pick up physical signals, that man has not yet identified. Experiments have also been carried out on animals (dogs and cats) that "know" when the owner is coming home, even at unusual times, and manifest this awareness with clearly explicit attitudes, such as standing by the door to welcome him as soon as he arrives.

The Feeling of Being Stared At

The feeling of being observed is quite common, and often it is also directional, that is, we perceive from where someone is observing us, so we can turn around in time to see that there is actually someone looking at us.

Actually very often it happened to ourselves that we were not observed but observers: for an unknown reason, we find ourselves looking at a person behind him or sideways, and suddenly this person turns around and meets his gaze with ours.

Evil Eye and Fascination

Malicious looks are a curiosity of human behavior that has always been known and has always been characterized by negativity. In the book *Evil Eye*, published in 1895 by Frederick Thomas Elworthy, it is argued that the eye has the power to emanate a force, and that when that force springs from the eye of envious or angry people, it infects the air.

In truth, Elworthy had already had worthy precursors. The belief in the evil eye during antiquity was supported by educated men such as Aristophanes, Plutarch and Heliodorus. Some detractors of Socrates claimed that this great philosopher possessed the evil eye.

We cannot speak of the evil eye without also mentioning its opposite, fascination, that is the ability to charm, to enchant with the look. More generally, fascination occurs when we allow ourselves to be positively involved in a show, a film, or something that attracts our attention, as can be the behavior of a person who "*fascinates us*".

Often the fascination is interpreted as the induction of a hypnotic state that serves to render impotent those who suffer it: the best known example in nature is the deadly fascination of the snake which, simply looking at its victim, immobilizes it and makes it unable to react.

Less tragically, we can say that, for example, the look of a puppy has the ability to fascinate us, as of course that of a child or loved one.

Figure 54. Princeton University, located in Princeton, New Jersey, is one of the largest universities in the Ivy League and is recognized as one of the most prestigious universities in the world. This is where the Global Consciousness Project is based.

27. The Global Consciousness Project

In the previous chapter, when presenting Dean Radin's experiments, I pointed out the surprising feature that emerged: the people subjected to the experiment perceived in advance the most shocking images appeerence and reacted, with measurable electrical signals, a few milliseconds before they were presented on the screen.

This phenomenon of early perception of unpleasant events has been recorded many times by many researchers.

The question someone might ask is: if it is possible to predict negative situations in advance by testing a person, would not it be possible, by testing the inhabitants of a city, to foresee negative events affecting that city?

By analogy, if negative events cause variations in the consciousness of human beings before they occur, would it not be possible, by testing the disturbances of vast sections of the population, to foresee negative events affecting entire nations or continents, or the whole planet?

Someone Tried. The Google Profile of Mood States (GPMOS)

A group of computer scientists and statisticians from the University of Indiana tried to feel the pulse of the collective mood by analyzing the moods of social networks, especially Twitter. To do this, they used a tool provided by *Google, the Google Profile of Mood States* or GPMOS. For eight months, from March to December 2008, scholars analyzed the moods recorded by the GPMOS, which are six: happiness, kindness, alertness, sureness, vitality and calmness.

They chose the periods in which the humoral state of calm prevailed, comparing it with the trend of the stock exchange, and they verified that, when the sample expressed this situation, the stock market closed without surprises with a forecast accuracy of 88 percent.

Someone Succeeded. The Global Consciousness Project (GCP)

Someone else has been involved in taking the pulse of the world in a more scientific way, with research still in progress that takes place within the project created in the 1980s by professors Robert Jahn and Brenda Dunne at Princeton University, the *Princeton Engineering Anomalies Research*.

This project was born when the various experimentations let us imagine that a collective global consciousness could really exist.

The studies consisted in verifying the psychokinetic effects on some small machines, of the electronic generators of random numbers, in technical terms RNG (*Random Number Generator*).

The numbers can be odd or even, so, as in the coin toss, after a reasonable number of tosses the result becomes flat (50-50). The people involved in the experiment were asked to try to influence the output of the numbers, and it was found that volunteers could actually generate interesting variations from the statistics.

Subsequently, it was verified that, by placing the RNG generators in a circumscribed environment, without anyone trying to influence them, the results were often different from those statistically predictable. It was concluded, therefore, that the RNG generators could be influenced by the moods of the people present within their range of action, without them voluntarily trying to do so, or even without their knowledge. There seemed to be a "community consciousness" that interacted with the generators, depending on the prevailing moods.

The experiment was called Mega Trial and continued with the involvement of two other research institutes, in Freiburg and Giessen in Germany.

In the following years, the interesting results and the greater sophistication of the technologies allowed to join to the project (which took the name of Global Consciousness Project) many other Institutes. Random number generators were placed practically all over the world, from Europe to the USA to Russia and then also in Japan, Brazil, China, South America, Australia and Africa. A total of 150 powerful RNG generators were installed.

These generators, still in operation, operate without the intervention of volunteers who try to influence them. They generate random numbers (one or zero) hundreds of times per second, and at the same time they formulate a prediction, then check if they have guessed correctly. According to the statistics, this should be done in 50% of cases.

But this does not always happen. Before significant collective events occur, generators match predictions much more often than the average, as if their ordinary functioning were influenced by events that will happen shortly thereafter.

In practice, the RNGs are able to register a significant change in the collective consciousness of communities within their sphere of influence, which is expressed in a greater number of exact predictions.

If the event can affect an entire continent, all generators on that continent are subjected to the same variations. As confirmation, huge peaks of exact forecasts were noticed a few hours before the attack on the Twin Towers in New York, on 11 September 2001.

These peaks continue to occur, locally or worldwide, in the hours preceding events that shortly afterwards will emotionally involve the populations within their range. It must be said, that not only negative events affect generators, but also positive ones, as happened at the Olympics opening ceremonies or other major international sporting events.

A feeling of great anxiety generates an exceptional increase
in the tools of the Global Consciousness Project.
This takes place 2 HOURS BEFORE the attack on the Twin Towers.

Figure 55. New York, September 11, 2001. The Global Consciousness Project's instrumentation is experiencing very high tension in the world's population two hours before the plane crashes into the first tower.

The only problem is that in most cases the instruments let us understand when humanity is in tension over something that has to happen, but it is not possible to establish how and where.

The Twin Towers

Perhaps, the most tragic event reported by the Global Consciousness Project was the one that took place in New York on September 11, 2001, with the attack on the Twin Towers and their collapse. Indeed, that day there was a series of four suicide attacks, which caused the death of about 3,000 people and the injury of another 6,000.

> "In examining the results of the analysis, we noticed that something unusual happened that day. On 11 September 2001, the curve of the graph underwent an incredible deviation in comparison with the other days examined. It turned out that this curve peaked about two hours before the hijacked plane crashed on the first of the Twin Towers, in New York, at 8.46 am local time.
>
> There is no easy answer to the question of why the curve peaked before the terrorist attack, although it does remind us of the data obtained from precognition experiments. What caused this big change?

Has the massive coherence achieved by the whole of minds, at world level, that day been reflected in the behavior of random number generators? It seems that things have gone this way. "

The Global Consciousness Project Today

In the years from 1980 to 2002, the project work continued under the direction of prof. Roger D. Nelson, who is now its Coordinator. The work of prof. Nelson, specialized in the study of consciousness and intention and in the role of mind in the physical world, integrates science and spirituality and provides a research model that focuses directly on numerous common experiences.

Nelson used RNG technology to study the effects of specific states of consciousness of human groups.

This today consists of a globally distributed network of RNGs, which continuously send data to a server located at Princeton University. The network has been designed to record indications of a global consciousness, that gets excited at major world events such as the beginning of wars, natural disasters, major sporting events and major celebrations such as New Year's Eve.

28. Conclusion. Is This the Right Time?

Since the dawn of civilization, humanity has always progressed in a non-linear way, although Darwinism has supported the theory of a gradual evolution of species, which progressively, through favorable mutations, proceeded in a path of adaptation to environmental conditions, the so-called *Phyletic gradualism*.

In 1972 two scholars, Stephen Jay Gould and Niles Eldredge proposed the *theory of punctuated equilibrium*. Based on the study of fossils, the two scientists contested the linear progressiveness of evolution, postulating that species remain stable for a long time and vary in short periods.

In practice, evolutionary changes take place relatively quickly, under the influence of environmental selective forces; these explosions of evolutionary variation would be interspersed with long periods of stability. The two scientists referred to the evolution of species on a biological level, but it is obvious that, as far as man is concerned, biological evolution is also accompanied by cultural evolution.

Indeed, we ourselves wondered in chapter 14, why man stood still for nearly four million years at the Stone Age, and then evolved with incredible speed over the last 12,000 years, up to the Computer Age.

In this regard, I supported the principle of the right moment: for everything there is a right moment to realize it, a moment in which the external objective conditions are propitious to make its realization easier.

The right moment manifests itself in a series of synchronicities that incite to pursue it, and stimulate the individual recipient, or an entire community, to embark on the adventure of change.

There are right moments for each of us, but above all there are right moments for the aggregations, the peoples, the whole of humanity. These are moments in which it becomes possible to make prodigious leaps in the awareness of new levels of culture and civilization, towards which to strive. In these cases, particularly powerful, insistent and repeated synchronicities push men to make the evolutionary leap.

I believe that we are experiencing one of these moments, that is, the passage from the civilization of matter to the civilization of matter united to the psyche. An incredible series of synchronicities is leading us towards this goal, an evolutionary stage of the mind that will not leave anything unchanged: if humanity has the courage to make this leap, nothing will be the same as before.

We have witnessed the development of quantum physics, the incredible discovery of the non-local reality in which entanglement occurs, the recent developments of the Global Consciousness Project, all of which coincide with the theories of psychotherapist Carl Gustav Jung on the collective unconscious, and above all his meeting and collaboration with the quantum physicist Wolfgang

Pauli. We could recall dozens and dozens of other coinciding events, such as Rupert Sheldrake's studies on morphic fields or the physicist David Bohm's theory on the implicate and explicate order. All these events, concentrated in the last fifty years, represent a tumultuous river of synchronicity that generates swirls capable of sweeping away centuries of beliefs no longer consistent with reality. Among these, first of all materialism, but also disbelief and tacit condemnation that relegated extrasensory perceptions to the role of "illusions and deceptions".

It is true that, in the general confusion, whole armies of dishonest have taken advantage of it to deceive the neighbor by selling alleged powers or abilities. This has resulted in discredit and denial of everything; as they say, the result has been to throw out the baby with the bathwater.

It is time to regain credibility to the phenomenon of extrasensory perceptions. We need to tackle the subject seriously. We must be aware that the language of extrasensory phenomena is still unknown to us. Now, however, we know that a grammar and a dictionary exist, and we can commit ourselves to studying them.

Timeline of Quantum Mechanics and Entanglement

1924
Wolfgang Pauli outlines the "Pauli exclusion principle", which states that no two identical fermions may occupy the same quantum state simultaneously.

1925
George Uhlenbeck and Samuel Goudsmit postulate the existence of electron spin.

Werner Heisenberg, Max Born, and Pascual Jordan develop the matrix mechanics formulation of Quantum Mechanics.

1926
Lewis coins the term photon in a letter to the scientific journal Nature, which he derives from the Greek word for light.

Oskar Klein and Walter Gordon state their relativistic quantum wave equation, later called the Klein–Gordon equation.

Dal 1926 al 1932
John von Neumann lays the mathematical foundations of Quantum Mechanics in terms of Hermitian operators on Hilbert spaces

1927
Werner Heisenberg formulates the quantum uncertainty principle.

Max Born develops the Copenhagen interpretation of the probabilistic nature of wavefunctions.

Chandrasekhara Venkata Raman studies optical.

Dirac states his relativistic electron quantum wave equation, the Dirac equation.

1930
Dirac hypothesizes the existence of the positron. His textbook *Principles of Quantum Mechanics* is published, becoming a standard reference book that is still used today.

Pauli suggests that, in addition to electrons and protons, atoms also contain an extremely light neutral particle which he calls the "neutron." Later it is determined that this particle is actually the almost massless neutrino.

1935

Einstein, Boris Podolsky, and Nathan Rosen describe the EPR paradox which challenges the completeness of quantum mechanics.

Schrödinger develops the Schrödinger's cat thought experiment. It illustrates what he saw as the problems of the Copenhagen interpretation of quantum mechanics if subatomic particles can be in two contradictory quantum states at once.

1948

Richard Feynman states the path integral formulation of quantum mechanics.

1961

Clauss Jönsson performs Young's double-slit experiment (1909) for the first time with particles other than photons by using electrons and with similar results, confirming that massive particles also behaved according to the wave–particle duality that is a fundamental principle of quantum field theory.

1963

Eugene P. Wigner lays the foundation for the theory of symmetries in quantum mechanics as well as for basic research into the structure of the atomic nucleus.

1964

John Stewart Bell puts forth Bell's theorem, which used testable inequality relations to show the flaws in the earlier Einstein–Podolsky–Rosen paradox and prove that no physical theory of local hidden variables can ever reproduce all of the predictions of quantum mechanics. This inaugurated the study of quantum entanglement, the phenomenon in which separate particles share the same quantum state despite being at a distance from each other.

1974

Pier Giorgio Merli performs Young's double-slit experiment (1909) using a single electron with similar results, confirming the existence of quantum fields for massive particles.

Dal 1980al 1982

Alain Aspect verifies experimentally the quantum entanglement hypothesis; his Bell test experiments provide strong evidence that a quantum event at one location can affect an event at another location without any obvious mechanism for communication between the two locations.

1992

Four-photon orbital angular momentum entanglement is confirmed at Leiden University, in the Netherlands.

2009

Aaron D. O'Connell invents the first quantum machine, applying quantum mechanics to a macroscopic object just large enough to be seen by the naked eye, which is able to vibrate a small amount and large amount simultaneously.

2011

On December 2, 2011, two diamond samples of millimeter size at room temperature, separated by a distance of about 15 cm, are entangled.

2014

Scientists transfer data by quantum teleportation over a distance of 10 feet with zero percent error rate.

Il 27 settembre 2014

A group of physicists announced that they had created a spin singlet with at least 500,000 rubidium atoms cooled to a temperature of 20 millionths of kelvin using quantum correlation.

2016

Aephraim Steinberg of University of Toronto, Canada, with some Canadian and Australian colleagues solves the surreal trajectories paradox of photons through the quantum entanglement phenomenon, thus confirming Bohm's hypothesis.

Da giugno 2017

A group of Chinese researchers have used satellite technology for the first time to generate and transmit entangled photons across a record distance of 1,200 kilometres on Earth.

Bibliography

Amir Dan Aczel, Entanglement. The greatest mystery of physics.
Barbour Julian, End of the time.
Barrow John David, From zero to infinity. The great story of Nothing.
Barrow John David, The numbers of the universe,
Barrow John David, Why is the world a mathematician?
Barrow John David, look Frank The anthropic principle.
Beitman Bernard, Messages from coincidences.
Cambray Joseph, Synchronicity. Nature and Psyche In a connected universe.
Cantalupi Tiziano, Santarcangelo Donato, Psychism and reality. .
Capra Fritjof, The Tao of physics.
John Cederquist, Coincidences They don't exist.
Cesati Cassin Marco, We're not here by chance.. The power of coincidences.
Subrahmanyan Chandrasekhar, Truth and Beauty. The reasons for aesthetics in science.
Chinnici Giorgio, Case Guard. The secret mechanisms of the quantum world
Chopra Deepak, Coincidences
Ford Kenneth, The world of Quanta. Quantum physics For everyone.
Gamow George, The Adventures of Mr. Tompkins.
Gamow George, Mr. Tompkins ' New World.
Goswami Arneb, Quantum Lighting Guide.
Greene Brian, The plot of the cosmos. Space,
Greene Brian, The hidden universes of parallel reality And the profound laws of the cosmos.
Greene Brian, The elegant universe. Superstrings, hidden dimensions and the pursuit of definitive theory.
Hawking Stephen The Universe in a nutshell.
Hawking Stephen The theory completely. Origin and destination Dell Universe.
Hawking Stephen The great history of the time.

Hawking Stephen Do Big Bang For black holes. A brief history of the universe.

Heckler, Richard, Coincidences.

Robert Hopke, Nothing happens by chance.

Joseph Frank, The power of coincidences.

Young Carl The analysis of Dreams. Archetypes of the unconscious. Synchronicity.

Young Carl Memories, DreamsReflections.

Kane Gordon, The Garden of Particles Elemental.

Shani Mani Quantum. From Einstein In Bohr, quantum theory, a new idea of reality..

Rei Hans, Christianity and Chinese religiosity.

Lederman Leon, Hill Christopher, Physical Quantum for Poets

Licata Ignazio, Watching the Sphinx.

Motterlini Matteo, Mental traps.

Peat David, Synchronicity. A union between the matter e Psyche.

Popper Karl, The Ego and your brain.

Radin Dean. Intertwined minds. Psychic phenomena explained by quantum physics.

Rhine Louisa, Psychokinesis. in mind Dominates matter..

Schumacher Ernst, A guide to the Perplexed, the B

Sheldrake Rupert, The illusions of Science.

Sheldrake Rupert, The mind Extended..

Michael Smith, Young and Shamanism.

Sparzani and Panepucci. (Curators) Young and Pauli. The original correspondence: The meeting between psyche and matter.

Henry Stapp Quantum theory and free will..

Michael Talbot, All is a. Feltrinelli

Teodorani Massimo, Bohm. The Physics of Infinity.

Teodorani Massimo, in mind Creative. From the physical universe to intelligent life.

Teodorani Massimo, The entanglement. The Weave In the quantum world: particles To consciousness.

Teodorani Massimo, Synchronicity. The link between physics and psyche. Da Pauli Young ' s Next In Chopra.

Teodorani Massimo, The Atom and the particles Elementary.

Seems Frank The physics of Immortality.

John White, The encounter between science and spirit..

Claudio Widmann, Synchronicity and coincidences Significant.

Claudio Widmann, Introduction to Synchronicity.